AF735488

Chemical Sensitivity and Sick-Building Syndrome

Chemical Sensitivity and Sick-Building Syndrome

Yukio Yanagisawa
Hiroshi Yoshino
Satoshi Ishikawa
Mikio Miyata

CRC Press
Taylor & Francis Group
Boca Raton London New York

CRC Press is an imprint of the
Taylor & Francis Group, an **informa** business

CRC Press
Taylor & Francis Group
6000 Broken Sound Parkway NW, Suite 300
Boca Raton, FL 33487-2742

© 2017 by Taylor & Francis Group, LLC
CRC Press is an imprint of Taylor & Francis Group, an Informa business

No claim to original U.S. Government works

Printed on acid-free paper
Version Date: 20160721

International Standard Book Number-13: 978-1-4665-5634-8 (Hardback)

This book contains information obtained from authentic and highly regarded sources. Reasonable efforts have been made to publish reliable data and information, but the author and publisher cannot assume responsibility for the validity of all materials or the consequences of their use. The authors and publishers have attempted to trace the copyright holders of all material reproduced in this publication and apologize to copyright holders if permission to publish in this form has not been obtained. If any copyright material has not been acknowledged please write and let us know so we may rectify in any future reprint.

Except as permitted under U.S. Copyright Law, no part of this book may be reprinted, reproduced, transmitted, or utilized in any form by any electronic, mechanical, or other means, now known or hereafter invented, including photocopying, microfilming, and recording, or in any information storage or retrieval system, without written permission from the publishers.

For permission to photocopy or use material electronically from this work, please access www.copyright.com (http://www.copyright.com/) or contact the Copyright Clearance Center, Inc. (CCC), 222 Rosewood Drive, Danvers, MA 01923, 978-750-8400. CCC is a not-for-profit organization that provides licenses and registration for a variety of users. For organizations that have been granted a photocopy license by the CCC, a separate system of payment has been arranged.

Trademark Notice: Product or corporate names may be trademarks or registered trademarks, and are used only for identification and explanation without intent to infringe.

Library of Congress Cataloging-in-Publication Data

Names: Yanagisawa, Yukio, author. | Yoshino, Hiroshi, author. | Ishikawa, Satoshi, author. | Miyata, Mikio, 1936- author.
Title: Chemical sensitivity and sick-building syndrome / Yukio Yanagisawa, Hiroshi Yoshino, Satoshi Ishikawa, Mikio Miyata.
Description: Boca Raton, FL : CRC Press, [2017] | Includes bibliographical references and index.
Identifiers: LCCN 2016031401| ISBN 9781466556348 (hardback : alk. paper) | ISBN 9781315374451 (ebk)
Subjects: | MESH: Multiple Chemical Sensitivity | Sick Building Syndrome | Air Pollution, Indoor--adverse effects
Classification: LCC RB152.6 | NLM WA 30.5 | DDC 615.9/02--dc23
LC record available at https://lccn.loc.gov/2016031401

Visit the Taylor & Francis Web site at
http://www.taylorandfrancis.com

and the CRC Press Web site at
http://www.crcpress.com

FSC www.fsc.org MIX Paper from responsible sources FSC® C013056

Printed and bound in Great Britain by
TJ International Ltd, Padstow, Cornwall

Contents

Preface xv
Acknowledgments xvii
Authors xix
Introduction xxi

Chapter 1 Present Status of Chemical Sensitivity 1

Satoshi Ishikawa

1.1 Background 1
1.1.1 Disease Registration of CS 2
1.1.2 CS Studies in Europe 3
1.1.3 CS Studies in Japan 3
1.1.4 Sick-Building/House Syndrome 3
1.2 Clinical Symptoms 3
1.3 Psychiatric and Social Aspects 4
1.4 Chemical Agents Implicated in CS 5
1.5 Why the Diagnosis of CS Is Difficult 8
1.5.1 Reasons for the Lack of Understanding of CS 8
Endnotes 8

Chapter 2 Effects of Chemical Sensitivity on Patients' Daily Lives 11

Satoshi Ishikawa

2.1 Effects of Chemical Sensitivity 11
2.2 Countermeasures 12
2.2.1 Making Your Home, Workplace, and School Environment Safe 12
2.2.2 Food and Digestive Issues 12
2.2.3 Caution in Use of Daily Products 12
2.2.4 Avoidance of Environmental Chemicals 13
2.2.5 Effects of Electromagnetic Waves 13
2.2.5.1 ES and International Classification of Disease 10 (ICD) 14
Endnotes 14

Chapter 3 Promoting Understanding of Chemical Sensitivity 17

Satoshi Ishikawa

3.1 Medical Insurance in Japan 17
3.2 Understanding of CS at the Judicial Level 17
3.3 Education 18
3.3.1 Specific Education of Medical Doctors 18
3.3.2 Specific Education for Dentists 18

3.3.3 Specific Education for the Healthcare Team 19
3.3.4 Patient Education for a Comfortable Daily Life 19
3.3.5 Education of Laypersons 19
Endnotes 20

Chapter 4 Medical Facts 21

Mikio Miyata, Kou Sakabe, and Satoshi Ishikawa
4.1 Acute Toxicology, Intermediate Syndrome 21
4.2 Chronic Toxicology 21
4.3 Animal Experiments Using Extremely Low Dosages............. 21
4.4 Research on the Mechanism of CS... 23
4.4.1 Detoxification 23
4.4.2 Nervous Sensitivity Acquisition 25
Endnotes 26

Chapter 5 Diagnosis 29

Satoshi Ishikawa and Mikio Miyata
5.1 Detailed and Careful Interview 29
5.2 Neuro-Ophthalmological and Neurological Examinations in Conjunction with Ordinary Clinical Examinations in Japan 29
5.2.1 Tracking Eye Movement Examination....................... 30
5.2.2 Pupillary Function... 31
5.2.3 Standing Ability 32
5.2.4 Contrast Sensitivity Examination of Higher Visual Centers 33
5.2.5 Accommodation Examination of Eyes....................... 33
5.2.6 Cerebral Function Examination 33
5.2.7 Immune Examination.. 34
5.2.8 Respiratory Function... 34
5.2.9 Peripheral Venous Blood Oxygen Concentration....... 34
5.3 Chemical Load Tests 34
5.3.1 Use of a Clean Room... 34
5.3.2 Determination of Causative Chemicals (Challenge Test)... 34
Endnotes 36

Chapter 6 Treatment of Patients with Chemical Sensitivity 37

Mikio Miyata and Satoshi Ishikawa
6.1 Avoidance of Chemical Substances... 37
6.1.1 Food... 37
6.1.2 Water 37
6.1.3 Air 38
6.1.3.1 Use of an Air Purifier 38

6.1.4 Clothes 39
6.1.5 Medicine 39
6.2 Detoxification 39
6.2.1 Spa, Hot Spring, Bath, Low-Temperature Sauna 40
6.2.2 Drug Treatment 40
6.2.2.1 Conjugation 40
6.2.2.2 Vitamins 40
6.2.2.3 Symptomatic Therapy 41
6.2.2.4 Minerals 41
6.2.2.5 EDTA, Penicillamine, PAM, and Others 42
6.3 Alternative Therapy 42
6.3.1 Prayer and Meditation in Zen Buddhism 42
6.3.2 Eastern Medicine and Alternative Medicine 43
6.4 Nutritional Instruction 43
Endnotes 44

Chapter 7 Chemical Sensitivity in Children 45

Kazuhiko Kakuta

7.1 General Symptoms 45
7.1.1 Formaldehyde Concentration and Symptoms 45
7.1.2 TVOC Concentration and Symptoms 46
7.1.3 Red Cell Cholinesterase 46
7.2 Effects of Indoor Chemicals on the Intelligence and Cognitive Function of Children 47
7.2.1 Evaluation of Intelligence 47
7.2.2 WISC-III Test Results and Concentration of Indoor Chemicals 47
7.2.3 Development of Sick-House Syndrome 49
7.3 Case Report of CS in a Young Boy Who Had Been Treated with Sumatriptan Succinate and Selective Serotonin Reuptake Inhibitors 50
Endnotes 51

Chapter 8 Multiple Chemical Sensitivity—Medical Aspects from Germany 53

Klaus-Dietrich Runow

8.1 Chemical Sensitivity 54
8.2 Biochemical Individuality 54
8.3 Randolph's Specific Adaptation Syndrome in Response to Environmental Stressors 57
8.4 Causes: Fire in the Body and the Brain 59
8.5 Diagnostic Program Recommended for CS Patients 59
8.5.1 Metabolic and Nutritional Status (Organic Acid Profiles), Toxins 59
8.5.2 Toxic Metals 60

8.5.3 Phthalates and Parabens Profile ... 61
8.5.3.1 Why Assess Phthalate and Paraben Levels? ... 61
8.5.3.2 Where Are Phthalates and Parabens Found? ... 61
8.5.4 Bisphenol A ... 62
8.5.5 Hidden Food Allergies ... 62
8.5.5.1 Gluten Sensitivity Can Cause Neurological and Autoimmune Disease ... 64
8.5.6 Stool and Digestive Analysis ... 64
8.5.6.1 Why Use Stool Analysis? ... 65
8.6 Therapy: The DNA Concept ... 65
8.7 HEPAR-TOX Detoxification ... 65
8.8 Ecological Architecture and Environmental Medicine ... 66
8.8.1 New Projects in Wolfhagen—The Historic City in the Land of the Brothers Grimm ... 66
Endnotes ... 67

Chapter 9 Emission Rate of Chemical Compounds in Building Products and Materials ... 69

Shin-ichi Tanabe
9.1 Introduction ... 69
9.2 Categories of Chemical Substances ... 69
9.3 Building Products and Materials ... 70
9.4 Emission Tests Using a Chamber ... 70
9.4.1 Small-Chamber Method ... 70
9.4.2 Large-Chamber Method ... 71
9.4.3 Desiccator Method ... 72
9.4.4 Passive Method ... 72
9.4.5 Microchamber Method for SVOCs ... 73
9.5 Labeling ... 75
9.5.1 Formaldehyde ... 75
9.5.2 VOCs ... 76
9.5.2.1 For VOC Standard ... 76
Endnotes ... 77

Chapter 10 Ventilation Strategies for Each Kind of Building and Statutory Regulations ... 79

Haruki Osawa and Masaki Tajima
10.1 History and Background of Indoor Air Pollution Measures in Japan ... 80
10.1.1 Before the Dawn of Measures against Sick Houses ... 80
10.1.2 Design Assumption and Strategy of Measures against the Indoor Air Pollution Problem ... 81

10.2 Statutory Regulations for Indoor Air Quality ... 83
10.2.1 Technical Standards in the Amended Building Standard Law ... 83
10.2.2 Regulation Regarding the Use of Building Materials Containing Formaldehyde and Installation of Ventilation ... 83
10.2.2.1 Restrictions on Interior Finishing ... 83
10.2.2.2 Mandatory Installation of Ventilation ... 83
10.2.2.3 Restrictions Concerning Attic and Adjacent Concealed Spaces ... 84
10.2.3 Technical Standards in the Housing Quality Assurance Act ... 84
10.2.4 Technical Standards of the Act on Maintenance of Sanitation in Buildings ... 85
10.2.5 Indoor Air Pollution Following the Amendment of the Building Standard Law in 2003 ... 87
10.3 Types and Features of Ventilation Systems ... 88
10.3.1 Types of Ventilation ... 88
10.3.2 Methods of Selecting a Mechanical Ventilation System ... 89
10.4 Trends in Technology Development and Energy-Saving Performance ... 90
10.4.1 Mechanical Ventilation System Utilizing Low-Power Input Motors ... 90
10.4.2 Demand-Controlled Ventilation ... 90
10.4.3 Mechanical Ventilation Systems with a Heat Exchanger ... 90
10.4.4 Hybrid Ventilation System ... 91
10.5 Important Points for Planning and Operation of Ventilation Systems ... 91
10.5.1 Planning of the Ventilation Path ... 91
10.5.2 Notes for Calculating Pressure Drops ... 92
10.5.3 Notes on Construction ... 92
10.5.4 Notes Concerning Regular Maintenance ... 92
10.5.5 Steps for Ventilation Planning ... 93
10.6 Ventilation Equipment in Large Buildings ... 93
10.6.1 Ventilation in Designated Buildings ... 93
10.6.2 Ventilation Systems Except Those Employed in Centrally Controlled HVAC Systems ... 94
Endnotes ... 95

Chapter 11 Ventilation, Air-Tightness, and Air Pollution ... 97

Hiroshi Yoshino and Rie Takaki

11.1 Outline of the Measurement Survey ... 97
11.1.1 Survey Period and Investigated Houses ... 97

11.1.2 Survey Points and Measurement Method..................97
11.2 Measurement Results..100
11.2.1 Air-Tightness of Houses..100
11.2.2 Ventilation Rate of the Ventilation System.............102
11.3 Comparison of Air-Tightness, Ventilation System, and Chemical Substance Concentrations.............................103
11.3.1 Comparison of Air-Tightness and Chemical Substance Concentrations..103
11.3.2 Comparison of Air Change Rate and Chemical Substance Concentrations..104
11.4 Conclusions..105
Endnotes...105

Chapter 12 Chemical Features of Indoor Pollutants and Current Regulations.... 107

Naohide Shinohara

12.1 Usage and Sources of Chemical Species Detected in Indoor Environments..107
12.1.1 Artificial Wooden Boards and Adhesives................107
12.1.2 Paint..110
12.1.3 Fungicides, Repellents, Mothballs, Air Fresheners, and Disinfectants.................................110
12.1.4 Plasticizers and Flame Retardants...........................111
12.1.5 Smoking and Burning Appliances...........................112
12.1.6 Secondary Generated Compounds...........................112
12.1.7 Contaminants in Outdoor Air..................................112
12.1.8 Unknown Pollutants...113
12.2 Determination of Pollutants from the View of Health Effects..113
12.2.1 Target Health Effects...113
12.2.1.1 Sick-House Syndrome..............................113
12.2.1.2 Asthma...114
12.2.1.3 Endocrine Disruption...............................114
12.2.1.4 Cancer..114
12.2.2 Field Survey of Indoor Pollution.............................114
12.3 Risk Assessment...118
12.3.1 What Is Risk Assessment?.......................................118
12.3.2 Hazard Assessment..118
12.3.3 Exposure Assessment...120
12.3.4 Risk Assessment...121
12.4 Enforcement and Effect of Regulations...............................123
12.4.1 Time Course of Regulations....................................123
12.4.2 Impact of Regulations..124
Endnotes...124

Chapter 13 Methods for Measurement of Indoor Pollution ... 127

Atsushi Mizukoshi

13.1 Methods for Measurement of Indoor Concentration ... 127
13.1.1 Integration Measurement ... 127
13.1.1.1 Active Method ... 128
13.1.1.2 Passive Method ... 128
13.1.1.3 Subject Compounds, Carbonyl Compounds, VOCs, TVOCs, SVOCs ... 128
13.1.2 Instant Measurement ... 130
13.1.2.1 Subject Compounds, TVOCs, Formaldehyde ... 130
13.1.2.2 On-Site Method of Measuring Source Intensity and Ventilation Rate ... 130
13.1.3 Combined Application of Integration Measurement and Instant Measurement ... 130
13.1.3.1 Peak Capture Method ... 131
13.2 Methods of Measuring the Emission Rate of Building Materials ... 131
13.2.1 Desiccator Method ... 131
13.2.2 Chamber Method ... 132
13.2.3 On-Site Measurement Method ... 133
13.2.3.1 Field and Laboratory Emission Cell ... 133
13.2.3.2 Passive Emission Colorimetric Sensor ... 133
13.2.3.3 Passive Flux Sampler ... 134
13.2.3.4 Advanced Diffusive Sampling Emission Cell ... 135
13.3 Methods of Measuring Personal Exposure Amount ... 135
13.3.1 Active Sampling–Passive Sampling Method ... 135
13.3.1.1 Search for Causative Compound of Chemical Sensitivity ... 135
13.3.2 Instant Measurement Method ... 136
13.3.2.1 TVOCs and HRV ... 136
Endnotes ... 136

Chapter 14 The Current Situation and Shift in Approaches to Indoor Air Pollution ... 139

Miyuki Noguchi

14.1 Framework of Laws and Regulations ... 139
14.1.1 Housing Quality Assurance Act ... 139
14.1.2 School Health and Safety Act ... 139
14.1.3 Act on Maintenance of Sanitation in Buildings ... 140
14.1.4 Revised Building Standards Act ... 140
14.1.5 Indoor Density Guideline Value ... 141

14.2 Trends after Determination of Guideline Values 142
14.2.1 Advantageous Effect of Formaldehyde Labeling..... 142
14.2.1.1 Formaldehyde Concentration in New Residential Housing................................. 142
14.2.2 Composition of VOCs 143
14.2.2.1 Significant Reduction of Concentrations of Guideline Substances.. 143
14.2.2.2 High Concentration Substances (TVOCs, Acetaldehyde, Aliphatic Compounds, etc.) 143
14.2.3 Widespread Use of Alternative Substances.............. 143
14.2.3.1 Low Formaldehyde Emission Adhesives ... 145
14.2.3.2 Water-Based Paint and Ink 145
14.2.3.3 Increased Use of Aliphatic Compounds ... 145
14.2.4 Temporal Change of Indoor TVOC Concentration... 146
14.2.5 Trends of Social Recognitions for Indoor Air Pollution: New and Remodeled Houses 146
14.2.5.1 Completion Checks by the Housing Quality Assurance Act.............................. 146
14.2.5.2 Widespread Indoor Air Pollution with Nonregulated Substances 146
14.2.5.3 Recognition That SHS Was Resolved....... 147
14.2.6 Change in Social Recognition of Indoor Air Pollution: Routine Countermeasures........................ 147
14.2.6.1 Air Cleaners... 148
14.2.6.2 Oxidative Decomposition 148
14.2.6.3 Secondary Pollutants 148
14.3 Required or Recommended Countermeasures...................... 148
14.3.1 Regulation of TVOC Concentration......................... 148
14.3.1.1 Enforcement of the Elimination Period.... 149
14.3.1.2 Information Sharing Among the Relevant Parties 149
14.3.2 Collecting Information from Medical Doctors 149
14.3.2.1 Selection Method for New Guideline Substances.. 149
14.4 Formulation of Counseling or Inquiry Systems for Patients and New Occupants... 150

Chapter 15 Investigation of Indoor Environments and Occupants' Health in Sick Houses 151

Hiroshi Yoshino, Sachiko Hojo, and Rie Takaki

15.1 Field Survey on Indoor Air Quality, Building Performance, and Occupants' Health of 62 Sick Houses 151
15.1.1 Description of the Investigation 151

15.1.1.1 Investigated Houses 151
15.1.1.2 Investigation of Building Performance and the Indoor Environment 152
15.1.1.3 Survey Questionnaire about Health Conditions 153
15.1.2 Results 153
15.1.2.1 Results of Chemical Substance Concentration Measurements 153
15.1.2.2 Relationship between Formaldehyde Concentration and Air-Tightness Measurement 153
15.1.2.3 Relationship between Formaldehyde Concentration and Air Change Rate 155
15.1.2.4 Sick-House Syndrome Based on the QEESI® Questionnaire 156
15.1.3 Chemical Substance Concentration and Occupants' Symptoms 156
15.1.3.1 Relationship between Chemical Substance Concentration and SHS 156
15.1.3.2 Relationship between Chemical Substance Concentration and Severity of Symptoms 157
15.1.4 Conclusion 157
15.2 Long-Term Observations Mainly for 30 Houses 157
15.2.1 Investigated Houses 159
15.2.2 Results of the Investigation 159
15.2.2.1 Relationship between Chemical Substance Concentration and Building Age 159
15.2.2.2 Yearly Changes of Occupants' Symptoms 160
15.2.3 Change of Symptoms and Chemical Substance Concentrations 161
15.2.3.1 Change of Symptoms in Response to Countermeasures against Mitigation of SHS 161
15.2.3.2 Decrease of Chemical Substance Concentration in Response to Countermeasures against Sick Houses 162
15.2.4 Conclusions 163
Endnotes 164

Chapter 16 Round Table Discussion 165

Yukio Yanagisawa

16.1 The Current Situation of Indoor Air Pollution 165
16.2 Medical Correlations in Japan 167

16.3 Government Actions ... 168
16.4 Patients' Compensation ... 169
16.5 Damp Buildings ... 172
16.6 Trade-Offs ... 173

Epilogue ... 175

Index ... 177

Preface

The impact of trace environmental chemicals on the human body, manifesting as sick-building or sick-house syndrome, has recently attracted social attention. The newly coined term for the diseases caused by trace environmental chemicals, "chemical sensitivity," currently used in modern society, may be truly secure. In particular, the impact of pesticides, which are used throughout the world, is becoming a significant problem. For example, the incidence of sexual maldevelopment and attention deficit hyperactivity disorder (ADHD) among children in the United States due to food contamination by toxic organophosphorous (OP) pesticides is becoming a major concern. Furthermore, in Brazil, hearing disorders caused by OP pesticides, discussed later, is another major problem. Among children in the farm area in Saku City, Japan, neurotoxicity, especially in the eyes, is an important concern reported as early as 1970–1980. After malathion was banned about nine years later and replaced with vamidothion, the symptoms ceased. Therefore, we concluded that the aforementioned diseases were caused by the OP pesticide malathion sprayed by the helicopters.

Serious neurological problems were also possibly induced by another OP pesticide, chlorpyriphos. Home use of chlorpyrifos was restricted in the United States in 2000, but it is widely used in agriculture, and is a serious risk to health and mental performance for people working and living in proximity to fields where it is used.

The mental impairment and birth defects caused by pesticides are autism and ADHD in 1987. Autism is observed in 1 out of 80 newborn babies in the United States and in 1 out of 50 newborn babies in Britain. Three cohort studies in the United States are tracking the long-term consequences of pesticide exposure on the developing brain during pregnancy and the early years of life. These studies have reported concerning results such as IQ deficits and ADHD-like behavioral problems.

Hearing loss caused by pesticides is an especially significant public health issue in Brazil. Pesticide sales skyrocketed from 2001 to 2008, making Brazil the world's leading consumer of poisons. A recent study aimed to assess whether pesticide exposure causes peripheral or central auditory disorders and thus focused on the importance of hearing tests in populations with acute or chronic exposure. I have previously reported the usage of OP pesticides in large amounts in Brazil.

The purpose of this book will be apparent from the ensuing discussion of the countless numbers of trace chemicals currently used for the convenience of present-day life that affect the human body.

Satoshi Ishikawa
Fellow American Academy of Environmental Medicine
Former Dean, Kitasato University School of Medicine, Japan

Acknowledgments

The editors and authors of each chapter would like to express their gratitude to the many researchers, listed below, who performed the studies and compiled the results described in this book.

Chapter 11: Koichi Ikeda (Nihon University), Atsuo Nozaki (Tohoku Bunka Gakuen University), Kazuhiko Kakuta (Kakuta Pedioatric Clinic), Sachiko Hojo (Shokei Gakuin University), Hideaki Yoshino (Environment Technology Solution Co, Ltd.), Kentaro Amano (Takenaka Corporation), Aki Nakamura (Japan Housing Finance Agency)

Chapter 15: Koichi Ikeda (Nihon University), Atsuo Nozaki (Tohoku Bunka Gakuen University), Kazuhiko Kakuta (Kakuta Pedioatric Clinic), Kentaro Amano (Takenaka Corporation), Aki Nakamura (Japan Housing Finance Agency), Naoya Ando (Taisei Corporation)

We express our heartfelt appreciation to Miwako Nyui, a secretary of Drs. Ishikawa and Miyata, and to the CRC Press editors of Leong Li-Ming, Lafoe Hilary, Kari Budyk, Judith Simon, Adel Rosario. Without their patience and support, we are confident that this book did not complete.

Authors

Yukio Yanagisawa, D. Eng. (Introduction, Chapters 12, 13, 14, 16). Yukio Yanagisawa is a professor emeritus at the University of Tokyo and currently the principal of Kaisei Academy Junior and Senior High School. His background is chemical engineering and public health. His research interests have been air environments, such as adverse health effects due to exposures to polluted air. He taught both at the Harvard School of Public Health and School of Frontier Sciences at the University of Tokyo. He was one of founders of the Society of Indoor Environment, Japan and a president or vice president between 1994 and 2008. He served as a vice president of Japan Society of Atmospheric Environment, an executive board member of The Japanese Society of Clinical Ecology, and Regional Environment Center for Eastern Europe in Hungary (REC). He is a Fellow of the International Society of Indoor Air Quality and Climate (ISIAQ).

Hiroshi Yoshino, D. Eng. (Chapters 9, 10, 11, 15, Epilogue). Hiroshi Yoshino is a professor emeritus of Tohoku University, and presently President-Appointed Extraordinary Professor at the university. He was the President of the Architectural Institute of Japan (AIJ). He has been involved in research subjects in building science such as indoor environment and energy conservation, ventilation and indoor air quality, occupants' health and indoor environment, and passive solar system performance. He is one of the contributors to the reports of the Intergovernmental Panel on Climate Change (IPCC), which was awarded the Nobel Peace Prize in 2007. He has received several awards, including the Architectural Institute of Japan in the area of journal papers and The Society of Heating, Air Conditioning and Sanitary Engineers of Japan (SHASE) Best Papers. He received the SHASE Japanese Uichi Inoue Memorial Award by the SHASE. He is an ASHRAE Fellow.

Satoshi Ishikawa, MD, PhD (Preface, Chapters 1, 2, 3, 4, 5, 6, 7, 8, 15). Satoshi Ishikawa is a professor emeritus at Kitasato University and a former dean of the School of Medicine, Kitasato University. His major research fields are neuro-ophthalmology and toxicology relating to clinical environmental Medicine. He served as a president of the Japanese Society of Clinical Environmental Medicine, a director of Chemical Sensitivity Study Group in Japan, a director of the , and a research director of the study group in 2000–2006 sponsored by Ministry of Health, Labor and Welfare. He is a Fellow of the American Academy of Environmental Medicine (AAEM) and a recipient of the Jonathan Forman Award.

Mikio Miyata, MD, PhD (Chapters 1, 2, 3, 4, 5, 6, 7, 8). Mikio Miyata is a professor emeritus at Kitasato University (Tokyo) and has worked at the chemical sensitivity special clinic in Tokyo. He is a former chairman of The Japanese Society of Clinical Ecology. He initially worked in the usual toxicology areas and later concentrated on the vital stimulation of a very small quantity chemicals. He proved that allergy strongly worsens by exposure to a very small quantity of chemicals. He then began to study chemical sensitivity. With Dr. Ishikawa he proved that patients with chemical sensitivity develop central and autonomic nervous system disturbances.

Introduction

In modern society, we are spending a large amount of time in indoor spaces. An indoor space is all the space separated from the open air, such as a house, a workplace, a means of transportation, a restaurant, a store, and so on. We who spend a large portion of time in indoor space inevitably are forced to breathe in the air in the places we occupy, even if the air is polluted. On the other hand, we can choose food and drinking water based on their quality and our preferences. It is not necessary to drink polluted water because clean bottled water can be carried anywhere. However, one cannot put clean air into a bottle and carry it. Therefore, contamination of indoor air may have serious consequences for our health. An adult normally takes in 2 kg of water and 2 kg of food per day, while taking in no less than 15 kg of air per day. Contamination of air, therefore, will have serious effects on our health.

Although contamination of indoor air could have serious effects on our health, in the past, residents, employees, doctors, and researchers seldom paid attention. Even if indoor air was contaminated, the outdoor clean air was immediately replaced by ventilation, because a house was not airtight. In modern society, however, a house is built airtight and people use an air conditioner to enjoy the comfortable temperature environment. The ventilation rate decreases. If temperature is adjusted by an air conditioner and the air of the indoor space is exchanged for the open air that differs in temperature, the energy consumption of an air conditioner will increase.

The large energy consumption induces a danger of bringing about global warming through an increase in the amount of discharged carbon dioxide from fossil fuel combustion as an energy source. To prevent the global environmental problems of global warming, it is desirable to cut down the energy consumption by increasing the air-tightness of a building and reducing the amount of ventilation. If the amount of ventilation is reduced, the environmental problem of room air pollution occurs because the polluted indoor air will not be exchanged for the fresh open air. We are faced with a trade-off in environmental problems between global warming and indoor air pollution.

Indoor air quality (IAQ) and its health effects are a very modern environmental issue; therefore, the present understanding of many people including doctors, builders, chemists, and, of course, occupants has not fully progressed to significant thinking regarding the sick-house syndrome and the chemical sensitivity induced by the polluted indoor air. This book intends to share the knowledge on this very modern environmental illness, that is, sick-house syndrome and chemical sensitivity, among all stakeholders. To prevent modern environmental illness, this book is written by experts in three academic fields: medicine, architecture, and chemistry.

There is a growing concern that human exposure to chemicals at levels once considered safe or presenting an insignificant risk could be harmful. Exposures in utero, during infancy, or over a lifetime are now suspected to have adverse biological effects on the development of the central nervous system development, affecting cognition, the immune system, and physical development as well. Disorders associated with chemical exposures are called by many names such as "sick-building syndrome," "sick-house syndrome," "sick-school syndrome," "multiple chemical sensitivity," "chemical sensitivity," "toxicant-induced loss of tolerance (TILT)," and "chronic fatigue syndrome." This indicates that we know these disorders cause much suffering not only in developed but also in less developed countries, but we do not know the specific biological mechanisms involved. The lack of clear biomarkers and time lag between initial exposures and ultimate symptoms make it technically, and increasingly politically, difficult to develop an extensive body of evidence needed to regulate many chemicals and industrial processes or to compensate the chemically injured. The emerging science associated with low-level chemical exposures requires that we examine both the way we think about chemicals and health and the solutions we devise to prevent chemically-caused injury. As scientists and citizens who assembled in the *International Symposium on Current Status of Indoor Air Pollution by Organic Compounds and Countermeasures for Healthy Housing*, January 13, 2001, in Tokyo, we appealed to everyone living in the twenty-first century to address these serious problems by applying the principles stated in the Right to Healthy Indoor Air by the World Health Organization (WHO) in 2000.

THE RIGHT TO HEALTHY INDOOR AIR (WHO 2000)

- P1. Under the principle of human right to health, everyone has the right to breathe healthy indoor air.
- P2. Under the principle of respect for autonomy ("self-determination"), everyone has the right to adequate information about potentially harmful exposures, and to be provided with effective means for controlling at least part of their indoor exposures.
- P3. Under the principle of non-maleficence ("doing no harm"), no agent at a concentration that exposes any occupant to an unnecessary health risk should be introduced into indoor air.
- P4. Under the principle of beneficence ("doing good"), all individuals, groups, and organizations associated with a building, whether private, public, or governmental, bear responsibility to advocate or work for acceptable air quality for the occupants.
- P5. Under the principle of social justice, the socio-economic status of occupants should have no bearing on their access to healthy indoor air, but health status may determine special needs for some groups.
- P6. Under the principle of accountability, all relevant organizations should establish explicit criteria for evaluating and assessing building air quality and its impacts on the health of the population and on the environment.
- P7. Under the precautionary principle, where there is a risk of harmful indoor air exposure, the presence of uncertainty shall not be used as a reason for postponing cost-effective measures to prevent such exposure.
- P8. Under the "polluter pays" principle, the polluter is accountable for any harm to health and/or welfare resulting from unhealthy indoor air exposure(s), and is responsible and accountable for correcting the condition.

P9. Under the principle of sustainability, health and environmental concerns cannot be separated, and the provision of healthy indoor air should not compromise global or local ecological integrity, or the rights of future generations.

Yukio Yanagisawa
Dr. of Engineering, Professor Emeritus, University of Tokyo;
Principal, Kaisei Academy Junior and Senior High School

1 Present Status of Chemical Sensitivity

Satoshi Ishikawa, PhD, MD, Professor Emeritus
Kitasato University

1.1 BACKGROUND

Various terms have been used to describe the hypersensitivity of the human body in response to contact with extremely small quantities of ambient chemicals. The designation currently most often used worldwide is chemical sensitivity (CS). The American pediatrician T. G. Randolph introduced this name around 1970, and it has been used since then. According to Randolph [1], a CS develops from an intertwining of three factors: (1) allergy, (2) nutrition, and (3) intoxication. According to a 2012 report by Bell and colleagues [2], CS affects about 5% of the US population. An examination of the Japanese population in 2000 showed that approximately 700,000 people out of about 120 million overall population are affected.

In this book, the general term CS is used. In Japan, research work on "the pathogenesis, diagnosis, treatment, and countermeasure of sick-house syndrome by a trace chemical substance" was performed by CS researchers in about 10 medical universities across the country under the support of the Ministry of Health, Labor, and Welfare during the 2000 to 2006 fiscal years [3,4]. At the work group, some additional novel names, for instance, low-dosage exposure sensitivity syndrome (LESS) and human hypersensitivity by chemicals in our environment, were proposed.

In 1987, M. R. Cullen, in the field of respiratory medicine, advocated the use of the term multiple chemical sensitivity (MCS). He examined workmen's accident patients with a focus on the respiratory system [5]. However, objective tests for diagnosis, such as genetic screening, a neurological examination, a brain imaging test, and an electrophysiology test, were not conducted. Therefore, in the United States, the word "multiple" was removed and the term "chemical sensitivity" is usually used in alignment with Randolph's designation. This terminology is used in Japan as well.

I also consider the name of "toxicant-induced loss of tolerance." Ashford, Miller, and others performed an epidemiological survey of CS among residents in New Jersey under governmental support in 1991 as stated by N. A. Ashford and C. S. Miller [6]. They reported in 1998 that "The serious health disturbance is caused by the low-dose chemicals exposure" and use the term "chemical sensitivity." However, they introduced chemical sensitivity using the term toxicant-induced loss of tolerance in the first (1991) and second editions (1998). A questionnaire for a medical examination, called Quick Environmental Exposure Sensitivity Inventory (QEESI), was introduced in this book. The Ishikawa translation of this questionnaire is used routinely in the medical examinations in Japan and is described in *Allegology & Immunology* [7]. In the United States, clinicians treating CS met and announced a consensus of CS in December, 1999. I mention this later. These reports represent the typical findings concerning CS.

Director W. J. Rea, who was also working in the CS area, and others of the environmental medical center in Dallas stated that though CS and many other terms exist, it cannot be said which is most accurate [8].

In Japan, administrative action followed from the findings of these CS work groups.

1.1.1 Disease Registration of CS

"Chemical sensitivity" was formally registered as a standard name in the disease roster in October 1, 2009. In Japan, CS has been formally recognized as an illness and therefore its treatment is covered by governmental health insurance.

Furthermore, the U.S. Washington State University environmental medicine department and Martin Pall in the biological chemistry field introduced chemical sensitivity in a standard toxicology textbook in October, 2009 [9]. His thinking is summarized as follows:

1. CS is a very common condition. In the United States, the number of CS patients, including latent patients, may be higher than that of people with diabetes. The epidemiological findings from nine nations are reported. According to Pall, 3.5% of the U.S. population is affected with CS.
2. CS is caused by poisonous environmental chemicals such as pesticides, organic solvents, volatile organic compounds, quicksilver, lead, carbon monoxide, hydrogen sulfide gas, and several other agents. CS induction by these substances was proven mostly in American universities. He stated that the *N*-glutamate receptor (NMDA receptor: *N*-methyl-D-aspartate receptor) is related to progression of the illness.
3. Work on genetic factors is presently ongoing in six universities in the world, primarily in the United States, Canada, and Germany. However, the principal causative factor in CS is chemicals.
4. Mode of pathogenesis and progression: A chemical generates peroxynitrite, which is a sort of active oxidization compound like superoxide. It causes neural sensitization and neurologic inflammation, and CS is induced.
5. Psychogenetic theory: In the United States, in the past 20 years a handful of people asserted a psychogenic origin of CS, in opposition to the scholars who earnestly investigated CS. They have no data and seriously interrupted and compromised researchers' work. The conclusion of Pall and others was, "Chemical sensitivity is absolutely not one of the abnormalities in psychophysiology. It is a sickness induced by the chemicals and is a sickness proven clearly by physiological and pathological etiology." Through Pall's work, the number of cases in which the verdict rules for the patient in CS litigation increased in Germany, northern Europe, and elsewhere after 2010.
6. Although recovery from CS was considered difficult, at present it gradually increases by treatment with antioxidants. After symptoms disappear, many

patients return to work completely. However, there have also been a few patients who tragically committed suicide, firmly believing that CS is an incurable illness. Evidence of a search for chemicals that can cause CS has been seen. An improvement of the surrounding environment, an appropriate medical and nutritional method based on pharmacology, saunas and hot spring treatments, massage, acupuncture, and general care of health have led to improvement of CS.

1.1.2 CS Studies in Europe

CS is a chronic disease. Headache, vertigo, and other unidentified, nonspecific complaints are repeated. The symptoms can develop in response to "a small quantity of chemical" in a manner different from that usually thought of when considering intoxication. They have stated that symptoms can occur when a person comes in contact with a trace amount of substance after previous exposure to a massive amount of chemicals, causing acute poisoning in some cases. Also, a chronic condition may develop after continuous exposure to trace chemicals in some cases.

1.1.3 CS Studies in Japan

In Japan, CS was observed around 1980, but government funding study by S. Ishikawa became available only in 1998 [10].

1.1.4 Sick-Building/House Syndrome

In Japan, the Building Standard Law relevant to Sick-Building or Sick-House syndrome was revised as a result of the findings of this work group in 2003. The indoor air quality guidelines created included about 13 kinds of chemicals, for example, formaldehyde, toluene, and organophosphorus compounds (OP compounds); chlorpylifos, diazinon, and other chemicals.

Very recently, a multiple CS in Korean adults was reported [11]. The new onset of CS is not decreasing because people still come in contact with OP pesticides, formaldehyde, and other chemicals in their surroundings.

Japan was faced with health disturbances caused by noxious environmental chemicals, for instance, methyl mercury and cadmium in the past. This book describes not only CS but also the environmental problems caused by other chemicals. In Asian nations, this kind of book hardly exists.

1.2 CLINICAL SYMPTOMS

As early as 1980, researchers conducting advanced studies in the United States and Japan noted the existence of an illness with allergy-like symptoms triggered by environmental chemicals. Among the causative agents mentioned were fumes from pesticides sprinkled in the garden, inside a house, in a park, and in a school and other buildings; exhaust gas from factories; the domestic town gas, perfume; tobacco smoke; formaldehyde gas; and others. Furthermore, CS research progressed

TABLE 1.1
Organ Abnormalities Caused by CS

No. of Cases	System Affected
45 (34%)	Central nervous system
32 (24%)	Sensory system, nose and throat system
26 (20%)	Digestive system
17 (13%)	Skin (e.g., eczema)
12 (9%)	Cardiovascular system

Note: Symptoms appears in multiple organ systems, especially in nervous system.

TABLE 1.2
Comparison of CS Cases in Japan and the United States

	Kitasato Clinical Environmental Center	Environmental Medical Center Dallas
Period	1996, 10.15–1997, 5.20	1994, 4.1–1995, 5.1
Average age	39.7 years	43.0 years
Age range	7–75 years	12–71 years
Female subjects	101 (75.4%)	61 (73.0%)
Male subjects	33 (24.3%)	23 (27.1%)

Note: Findings of CS are internationally similar.

in advanced countries and the results emphasize that it is an illness that causes abnormalities in many systems, such as respiratory, digestive, sensory, nervous, musculoskeletal, endocrine, and immune, and somatic symptoms generally (Table 1.1). An Ad Hoc Committee of the Canadian government conducted a detailed test of 132 adult cases diagnosed with CS in 1985. This was introduced by P. R. Gibson [12] and a comparative study of CS was conducted.

Table 1.2 shows a comparison between Japan and US CS cases clinically reviewed by W. Rea, with whom we collaborated [8].

It is clear that the findings for Japanese and US patients are similar.

1.3 PSYCHIATRIC AND SOCIAL ASPECTS

In numerous patients, CS had been diagnosed as a spiritual or psychosomatic illness. In cases in which the cause of CS is clearly exposure to certain chemicals, the problem of misdiagnosis is easily avoided. When the causative agent of the pathogenesis is not definite, when a patient has shown a predisposition to sensitivity, and when a patient's symptoms are extensive, the patient tends to receive a psychiatric diagnosis. The diagnosis of CS is more difficult when a patient has demonstrated psychosomatic disorders, panic disorder, phobia, and somatization disorders. In addition, when electromagnetic hypersensitivity is combined with CS, a medical diagnosis becomes much more complex.

To pursue a CS case in the judicial system, not only a patient's subjective complaint but also proof of abnormal findings by an objective test is required. Many chemical companies and experts retained by them claim strongly that CS is a spiritual and a psychological disease. These were typical opponents in the case of Minamata disease due to methyl mercury poisoning caused by eating fish contaminated with methyl mercury. Even now, the sequelae are ongoing and the disputes with the company, Chisso corporation of Japan, have not been resolved.

1.4 CHEMICAL AGENTS IMPLICATED IN CS

A disastrous accident at the Fukushima nuclear power plant in Japan occurred on March 11, 2011 after a huge earthquake and tidal wave (tsunami). Contamination of water was a major problem for the Japanese government and the Tokyo Electric Power Corporation. Although findings related to resulting water contamination were not complete, they emphasized to the public that "a quantity of this amount of exposure of both external contamination and internal contamination to the body is seldom problematic for humans in the extremely low dosage of the leakage into the water. Therefore, extreme unease is not needed. Don't believe rumors." We are not sure whether a detailed analysis of the water has been conducted.

The main chemical cause of CS seems to be OP, even with exposure to low dosages. How do we come into contact with OPs in our daily lives? They are found in common pesticides that are used extensively in public places—shopping malls, restaurants, schools, parks, hospitals, and even on trains—to get rid of cockroaches and control infectious diseases. At home, OPs are found in termite control sprays and tick-control sheets placed under tatami mats, and so forth. They are also major ingredients of outdoor insecticides and herbicides massively used in agricultural areas and in public and private gardens. OPs vaporize into the atmosphere, which means everyone inhales them without knowing.

In chronic OP intoxication, symptoms vary from patient to patient, but the most common early symptoms include eye problems such as defocusing and deteriorating vision, oculomotor involvement, and narrowing visual fields and autonomic nervous symptoms such as severe fatigue, muscle pain, headache, nausea vomiting, and dizziness. Next-stage symptoms include mental or emotional disorders, such as depression, emotional instability, lapses in memory, and sleeping difficulties.

Why do these conditions occur? Simply stated, OPs inhibit the activity of various enzymes and disrupt the natural mental biorhythm, distorting the neurological function of the brain. If those enzymes are repeatedly inhibited, the conditions worsen, beginning the onset of symptoms of CS caused by damage to the autonomic nervous system.

Where enzyme inhibition is concerned, researchers have traditionally focused almost entirely on the neurotoxic effect of acetylcholinesterase (AchE) inhibition. However, recent studies have confirmed the inhibition of various types of enzymes including fatty acid amide hydrolase (FAAH) that metabolize important signaling substances with an endogenous cannabinoid that regulates brain functions. These studies have shed considerable light on the explanation of the toxic nature of OPs.

The extent of enzyme inhibition varies greatly from individual to individual depending on the person's genetic makeup. This means that even if a person is not affected, someone else could be damaged by OPs the person may be using. Children in particular need special protection, because the development of their mental and neurological functions is extremely sensitive to OPs.

Ishikawa examined 71 new cases of schoolchildren with visual disturbance starting in 1966. Most of them were children of farmers in Saku City. A helicopter spray of OP pesticide (Malathion® 3%) was carried out several times a year from 1961 until 1969. The chief complaints of patients were reduction of vision in both eyes with bilateral visual field narrowing and poor pupil response to light and disturbed eye movement. Higher levels of autonomic nerve dysfunction accompanied with reduced motivation were seen in adult farmers. Symptoms in adults were similar to the symptoms and signs of CS patients described by Rea in the United States [8]. A comparison of children in the Saku case with healthy children in Tokyo showed that serum and the red cell cholinesterase (ChE) was reduced in 35% of the 71 cases. Liver function tests were abnormal as well. Neuro-ophthalmological tests revealed abnormalities in light reactions of pupils and smooth pursuit eye movement (especially vertical eye movement); disturbed body balance was demonstrated by a positive Romberg test.

Pralidoximemethiodide (PAM), a specific dephosphorylation agent, had been used together with atropine, an anticholinergic drug. After administration of pulifinium bromide, an anticholinergic agent, and glutathione the aforementioned neural and visual system symptoms improved. From these results, aerial application of the Malathion was concluded as the cause of the disease. The city administration stopped the Malathion spray in 1969. Furthermore, the country reduced the frequency of an aerial application. As a result, de novo patient onset disappeared. Comparison data for 71 patients and normal children (Tokyo) are shown in Table 1.3.

I presented these data at an International Association of Eye Research symposium in Paris [13,14]. Taking advantage of this opportunity, a large-scale collaborative supplementary examination work was started at the National Institutes of Health (NIH) in the United States, where in 1994 specialists announced that Saku disease was produced by chronic OP intoxication [15]. Jonathan Forman Prize in USA was awarded to S. Ishikawa in 1996 for the works of chronic OP intoxication.

As shown in Table 1.3, the Saku patients experienced an excessive tonic state of the parasympathetic nervous system (cholinergic state). Acetylcholine accumulated at the nerve ending site, causing a muscarinic and nicotinic intoxication [16]. Although the ChE level was lower in the patient group than in the control, the reduction was seen in only 30% of patients and the remaining 70% of patients had normal values. Consequently, OP intoxication assessed by ChE value did not become conclusive for the diagnosis. The occurrence of OP intoxication is still seen in the world. Countries in Central America, such as Cuba, have had cases similar to those in Saku disease.

A new systematic review of studies of pesticide exposure on the auditory system identified 143 studies on the topic. All articles concluded that pesticide exposure is "ototoxic" and leads to hearing loss [17]. There was almost no report of research of a hearing disorder and an OP pesticide in the past. Insecticide use is prevalent in Japan,

TABLE 1.3
Patients' Complaints at Saku Area and the Controls at Tokyo City (Rate of Appearance in Percent)

Complaint	71 Patients at Saku	100 Controls at Tokyo
Headache	6	3
Hyperperspiration	14	6
Motion sickness	14	6
Polydipsia	20	7
Dizziness	2	0
Palpitations	1	0
Hypotension	5	0
Hypertension	0	0
Hyperlacrimation	1	0
Nausea and vomiting	2	1
Constipation after diarrhea	6	1
Findings		
Reduced vision with no improvement by correction	100	7
Narrowing of the visual fields in both eye	95	6
Vertical astigmatism over 1.5D difference in horizontal and vertical meridians	88	5
Impaired smooth pursuit of eye movement	57	38
Pupillary abnormality, miosis, iridoplegia, delayed latency followed by the light reaction	52	34
Electroretinography (ERG)		
Supernormal type	33	3
Subnormal type	10	0
Nonresponse type	2	0
Normal type	55	93
Cholinesterase activity in serum (Michael's method)	0.8 ± 0.11*	1.02 ± 0.14
Cholinesterase activity in red cells (Ellman's method)	0.47 ± 0.43	1.88 ± 0.22**
	*$p < 0.05$	**$p < 0.01$

Note: Children at Saku area exposed chronically to OP show more complains of nervous system and more neurologically abnormal findings compared with those living at Tokyo. Both cholinesterase activities in serum and red cells of the children in Saku area are lower than the values of children in Tokyo.

and at present, Brazil leads the world in the amount of pesticide used in cultivated acreage. Full data for insecticide application in China have not been reported. Global awareness has been raised as a result of Brazil's concern about hearing disorders and the partimutism that progresses after exposure to organophosphorus pesticides.

Even if a patient complained of headache, vertigo, nausea, diarrhea, and so forth, it was often noted that "a specific symptom is missing." It has further been assumed that a number of CS patients had been neglected by physicians in Japan from 1970 to 1985.

1.5 WHY THE DIAGNOSIS OF CS IS DIFFICULT

At present, a number of people are still affected with OP intoxication. M. Abdollahi and others described cases of CS intoxication caused by OPs [18]. An intermediate syndrome was seen after 24–96 hours of onset in acute OP intoxication. Patients presented with whole-body tiredness, myalgia, and autonomic nerve dysfunction. The conclusive factors in the objective medical examination were abnormalities in the electromyogram (EMG). An amplitude reduction (waning phenomenon) of the muscular contraction caused by electronic stimulation of the radial and the ulnar nerves was detected. These are typical findings in the intoxication syndrome caused by OP pesticides, called organophosphate-induced intermediate syndrome. An intense nicotinic and muscarinic action appears in acute OP intoxication. In subacute intoxication, these symptoms are mild. Headache, vertigo, nausea, reduction in concentration, loss of muscle strength, and so forth will appear. In chronic intoxication, one has to look for a contact point with the OP agents. The physician who treats these patients should prescribe a cholinolytic drug such as atropine and an antidote such as glutathione for chronic patients. The improvement of symptoms by antidotes is called a therapeutic diagnosis.

1.5.1 Reasons for the Lack of Understanding of CS

For clinicians who were educated in classic toxicology, the diagnosis of acute poisoning is easy, but if they have not received a modern toxicological education, especially of chronic toxicity, they tend to object to the recognition of the name CS and to the physicians who diagnose CS.

In Japan, a strict regulation of the chemicals used in architecture was enacted in 2003. A standard value was assigned to 13 substances including formaldehyde. In Japan, the use of chlorpyrifos spraying at rice fields was forbidden. As a result of these policies, the number of patient onsets of sick-building syndrome (in Japan, called sick-house syndrome) might decrease. However, sick-building syndrome can still be caused by low-dosage chemicals other than the regulated 13 substances.

Moreover, regardless of the regulation of architecture, the number of CS cases generated by pesticides or herbicides sprinkled on the ground, yard, and so forth is not necessarily decreasing.

ENDNOTES

1. Randolph TG. 1987. *Environmental medicine: Beginnings and bibliographies of clinical ecology.* Clinical Ecology Publications.
2. Bell IR, Baldwin CM, Fernandez M, Schwartz GE. 1993. Neural sensitization model of multiple chemical sensitivity: Overview of theory and empirical evidence. *Toxicol Indust Health* 14:295–304.
3. Committee on Sick House Syndrome. 2000–2002 financial year. Supported by Ministry of Health, Labor and Welfare, Japan. Printed in 2001–2003. Chairman Ishikawa S.
4. Committee on Trace Amount of Chemicals and Sick House Syndrome. 2000–2002 financial year. Supported by Ministry of Health, Labor and Welfare, Japan. Printed in 2001–2003.

5. Cullen MR. 1987. Workers with multiple chemical sensitivity. *Occup Med* 2:655–661.
6. Ashford NA, Miller CS. 1996. Chemical exposures, low-level chemical sensitivity: Current perspectives. *Int Arch Occup Environ Health* 68:367–376.
7. Ishikawa S. 1999. Chemical sensitivity—A diagnostic criteria and a medical examination. *Allegol Immunol* 6:990–998.
8. Rea WJ. 1997. *Chemical sensitivity*, Vols. I–IV: *Tools of diagnosis and methods of treatment*. Boca Raton, FL: Lewis.
9. Pall M. 2009. Multiple chemical sensitivity: Toxicological questions and mechanisms. In Ballantyne B, Marrs TC, Syversen T (eds.), *General and applied toxicology*, 3rd ed., Vol. 4, pp. 2303–2354. Chichester, UK: John Wiley & Sons.
10. Ishikawa S, Miyata M, Namba T et al. 1998. Diagnostic criteria of chemical sensitivity. [In Japanese.] *Japan Medical Journal* 3857:25–29.
11. Jeong I, Kim I, Park HJ. 2014. Allergic diseases and multiple chemical sensitivity in Korea. *Asthma Immunol Res* 6:409–414.
12. Gibson PR. 2000. *Multiple chemical sensitivity: A survival guide*. Oakland, CA: New Harbinger Publications.
13. Ishikawa S, Miyata M. 1980. Development of myopia following chronic organophosphate pesticide intoxication: An epidemiological and experimental study. In Merigan WH, Weiss B (eds.), *Neurotoxicity of the visual system*, pp. 233–254. New York: Raven Press.
14. Ishikawa S. 1976. *22nd Concilium Ophthalmologium* Paris 1974, *ACTA* Vol. 2, pp. 1021–1025. Paris: Masson.
15. Dementi B. 1994. Ocular effects of organophosphates: A historical perspective of Saku disease. *J Applied Toxicol* 14:119–129.
16. Jaga K, Dharmani C. 2006. Ocular toxicity from pesticide exposure: A recent review. *Environ Health Prevent Med* 11:102–107.
17. Kos MI, Hoshino AC, Asmus CI et al. 2013. Peripheral and central auditory effects of pesticide exposure: A systematic review *Cademos de Saude Publica* 29:1491–1506.
18. Shadboorestan A, Vardamjari HM, Abdollahi M et al. 2016. A systematic review on human exposure to organophosphorus pesticide in Iran. *J Environ Sci Health C Environ Carcinog Ecotoxicol Rev* 34:187–203.

2 Effects of Chemical Sensitivity on Patients' Daily Lives

Satoshi Ishikawa, PhD, MD, Professor Emeritus
Kitasato University

2.1 EFFECTS OF CHEMICAL SENSITIVITY

A patient who reacts to everyday products such as spices, detergents, tobacco, organic solvents, and so forth faces a risk of discord in marital and family relationships. People may react to chemical substances in a workplace, and in products the other personnel are using, as well as to new office fixtures and building materials. These individuals find it difficult to work in an office. They can use only additive-free foods, clothing, and home products. Income could be compromised while expenses increase, and such persons can become destitute.

Some people are born with a chemical sensitivity (CS) and others develop an allergy. Many patients acquire CS from a sick house. There are also many patients who repeatedly move into newly built dwellings and show symptoms. CS may in addition occur as a result of medical or dental care.

A patient first must become skillful in avoiding exposure to chemical substances, after which symptoms will improve gradually. However, the kind of chemical substance to which the patient reacts increases and the symptoms that appear at the time of chemical substance exposure also increase during a course of sickness. This phenomenon is called spreading. A patient will be frightened. If a CS patient's awareness and avoidance of chemical substance exposure progresses, it will be a natural reaction to the diagnosis of CS. However, I am undecided whether one should inform a patient that symptoms may increase in the future because such an explanation may amplify a patient's uneasiness. A CS patient's fear of insecurity is strong [1]. Seventy percent of CS patients mostly recover after medical treatment. Of course, the recovery means that a patient needs to carefully avoid chemical substance exposure in our clinic, environmental. What factors lead to abatement of symptoms?

This theme is directly linked to medical treatment. Physically it consists of detoxification and psychologically of removing unease and promoting a stable mental outlook based on the medical treatment. Sensitivity often reinforces uneasiness, and if it progresses, it will proceed even to a delusion of injury. Patients' mental status can be fragile; they should not feel confined to a room with a feeling of gloom (Figure 2.1).

Chemical sensitivity (CS)

↓

Anxiety

↓

Anxiety neurosis

↓

Delusion of injury

FIGURE 2.1 The mental state of patients with chemical sensitivity is fragile.

2.2 COUNTERMEASURES

2.2.1 Making Your Home, Workplace, and School Environment Safe

All food, clothing, and shelter must be considered. Exclusion of a air pollution/chemical substance is especially important. Identification of a sick building, a sick house, and a sick school shows the importance of the air in each case. Therefore, this book is authored by a specialist in construction architecture and a specialist in environmental medicine. An average adult inhales 15–20 kg of air every day. A food contamination substance passes through the liver, which is a gateway for detoxification after gastroenteric absorption, and a considerable portion is decomposed. An air pollution chemical substance is absorbed without a barrier to prevent its entry into blood; air pollution is therefore dangerous.

2.2.2 Food and Digestive Issues

More than 70% of people with chemical sensitivity had common allergies to dust, pollen, and food as well as food additives. Symptoms of allergy appear to center on the immune system. As for CS, symptoms appear to focus on the nervous system. The majority of serious CS illnesses are complications of a food allergy. Nutritional treatment is a large pillar in management of a CS without a specific medicine. In Japan, more than 20 g of a synthetic food additive is taken in by each person every day. The phosphate absorbed reacts with the chloric acid of gastric juice and will become phosphoric acid and a chloride. The metabolism of calcium, which has a leading role in intracellular communication, will be disrupted.

The chemical substance to which a CS patient reacts has, fundamentally, a deleterious effect on the body.

2.2.3 Caution in Use of Daily Products

Products in daily use contain many hazardous substances. Vinyl chloride monomer and plasticizers are volatilized from vinyl chloride, which is used in medical supplies such as intravenous drip tubes and inhalation tubes. Some anesthesiologists have a misunderstanding of oxygen as a gas with an odor. A doctor may be unaware of continuously exposing a patient to a toxic substance. I recommend that doctors at least use tubes that are free of (poly)vinyl chloride (PVC) and wish to emphasize again that the substance to which a patient reacts is fundamentally hazardous.

A starting point in the chemical industry is the production of sodium hydroxide from sodium chloride. The industry is taking pains over the consumption of the chlorine of the byproduct. There is nothing gentle in the effect on the body of a chlorinated organic compound. The simple polyethylene film of wrapping film is unlikely to cause harm. Some of the everyday products containing chemical substances that should be avoided include fire retardants, floor wax, car wax, synthetic perfume, synthetic detergent, hand soap, fiber softening agent, a sweat control spray, perfume, a deodorizer, antibacterial medicine, stationery, personal computer, the toner of copy machine, rubber commodity, leather article, shoe polish, shampoo, some tooth brushing agents, moth balls, and so forth. Among specific examples are melamine and formaldehyde from the melamine resin of tableware, antiseptics from half-split chopsticks, and heavy metal coloring flowing out from a picture adhering on the surface of pottery. Another is tobacco. A bad smell emanates from a smoker's exhalation, clothes, and hair. A smoker poses a serious danger for patients with CS as well as others.

2.2.4 Avoidance of Environmental Chemicals

Although pesticides, herbicides, and fungicides are generally called agricultural chemicals, such chemical substances that kill living things are not good for humans. Use of these in agriculture is unavoidable. However, these compounds should not be used at home. The no. 1 cause of CS is said to be insecticides. Extermination of termites often causes sick-house syndrome [2].

Paint, glue, formaldehyde, toluene, styrene, and phenol are always used at the time of new building construction. In reconstruction of a building, or in modification and waterproofing work, they are unavoidable. Such materials also volatilize from large furniture. When it is difficult to avoid such chemical pollution exposure, people need temporary refuge.

2.2.5 Effects of Electromagnetic Waves

Electromagnetic waves may have caused various kinds of physical difficulty. The energy absorbed affects the body as an inevitable consequence of the law of conservation of energy. Numerous effects of electromagnetic waves on a living body have been reported, such as a visual and genitourinary system abnormalities, miscarriage, stillbirth, inborn errors of metabolism, developmental disorders, premature aging, cancer, leukemia, brain tumor, blood–brain barrier disorder, alterations in neurotransmitters, Alzheimer's disease, amyotrophic lateral sclerosis, Parkinson's disease, sleep disorder, depression, anxiety, epilepsy, schizophrenia, behavior disorder, recognition disorder, arteriosclerosis, high blood pressure, myocardial infarction, diabetes, immune system disorders, and so forth [3].

Patients with CS often complain of electric sensitivity (ES). It is generally said that in Japan and the United States ES appears in 60% of CS patients. The subjective symptoms are as follows: neurological symptoms such as headache, difficulty in concentrating, sleep disorder, fatigue, digestive symptoms such as diarrhea or constipation, stomachache, and loss of appetite; skin symptoms such as burning sensation, prick pain, rash, itchiness; eye symptoms such as eye pain, dry eye, blepharospasm, blur; ear, nose, and throat symptoms such as dizziness and nausea; and motor organ symptoms such as muscular pain, neuralgia, and others.

Electromagnetic waves, as well as organophosphorus (OP) compounds, cause the blood–brain barrier to open, and a toxic substance can easily enter the brain. As mentioned previously, calcium has a leading role in intracellular communication [3]. Electromagnetic waves change the calcium metabolism in a cell, causing abnormalities in brain neurotransmitters such as acetylcholine, noradrenaline, serotonin, gamma-aminobutyric acid (GABA), glutamate, and dopamine. Electromagnetic waves further induce oxidative reactions. K. Runow created a clinic with very little electromagnetic wave exposure in Germany in 1987.

Ishikawa applied a pupil test as one of the autonomic nerve examinations for CS and ES patients [4,5]. He objectively discovered abnormalities of both sympathetic and parasympathetic nerve functions in patients.

Medical treatments of ES are substitution therapy consisting mainly of application of functional therapy. A reduction of electromagnetic wave exposure and anti-oxidization treatment, minerals (Ca, Mg, Se) and vitamins are fundamental treatments. In parallel with this, a bathing medical treatment in the hot spring at the Emstal area has performed well as the orderly plan. These methods have constituted a successful and effective medical treatment since 1980 [6].

2.2.5.1 ES and International Classification of Disease 10 (ICD)

Studies on ES have progressed in Japan [7,8] but the name ES was not formally accepted in the medical and welfare field in Japan.

Austrian university faculty members, medical association member physicians, researchers, and electromagnetic wave specialists gathered on March 3, 2012, and wrote a consensus paper of the Austrian Medical Association on electromagnetic field syndrome, (EMF) [9]. They recommended that the code Z58.4 (Exposure to radiation) under ICD-10 be used for EMF syndrome as T65.9 for CS. ES is completely formally accepted as a diagnosis by the Austrian Medical Association. A series of questionnaires is the fundamental procedure for diagnosis. The questionnaires are nearly similar in Japan [8].

ENDNOTES

1. Tonori H, Aizawa Y, Ojima M et al. 2001. Anxiety and depression in multiple-chemical sensitivity. *Tohoku J Exp Clin Med* 193:115–121.
2. Itayama K, Kameya T et al. 2006. Characteristics and causes of self-reported idiopathic environmental intolerances (multiple chemical sensitivity). *Chems Bio Integrated Management* 2:178–191.
3. Sakabe K, Hane K, Miyata M. 2012. Living body and electromagnetic wave 2012. Maruzen Publishing Co., Ltd. Tokyo.
4. Ishikawa S. 1996. About ocular manisfestation by environmental pollution etc., especially on the visual toxicity of an organophosphorous pesticide. *Jpn Ophthalmol Soc* 100:417–432.

5. Ishikawa S, Miyata M, Sakabe K et al. 2006. 7 cases of electromagnetic hypersensitivity. Annual report of trace chemicals and sickhouse syndrome in 2005 financial report. Supported by Ministry of Health, Labor and Welfare, Japan 30–41.
6. Runow Klaus-Dietricj. 2011. Wenn Gifte auf die Nerven gehen. IFU Bad Emstal, Germany: Verlag Suedwest. E-mail: info@ifu-Wolfh agen.de.
7. Hojo S. 2016. Electromagnetic field as novel health risk factors—Issues identified from an epidemiologic study on electromagnetic hypersensitivity. *Jpn J Clin Ecol* 25(2), in press.
8. Committee on Trace Amount of Chemicals and Sick House Syndrome. 2006. 2005 financial year. (pp. 30–48). Ishikawa S, chairman. Supported by Ministry of Health, Labor and Welfare, Japan.
9. Guideline of the Austrian Medical Association for the diagnosis and treatment of EMF-related health problems and illnesses (EMF syndrome): Consensus paper of the Austrian Medical Association's EMF Working Group (AG-EMF), 2012.

3 Promoting Understanding of Chemical Sensitivity

Satoshi Ishikawa, PhD, MD, Professor Emeritus
Kitasato University

3.1 MEDICAL INSURANCE IN JAPAN

In Japan in a notification from the Ministry of Health, Labor, and Welfare, the name of the disease "chemical sensitivity (CS)" was newly exhibited on October 1, 2009 (MEDIS-DC, concerning the standard name in a disease roster), and was formally registered as the name of a disease for insurance purposes. The name may be applied to everyday medical situations. This determination is likely to be good news for patients who are concerned about this illness.

After CS was adopted as the health insurance name of a disease, the official support organization of patients, such as medical treatment, compensation for absence from work, and disability pension, improved greatly. Private insurance compensation payment is going smoothly, without CS being ignored by insurance companies. Of course, for CS patients, future problems, such as an escape facility at the time of neighboring air pollution, nursing care, and entrance equipment for elderly people, remain.

In Japan at present, authorization of CS as the cause of a workman's accident is still a somewhat difficult problem. The judgment committee determines the cause of the accident. The committees have not advanced beyond the idea of authorization for a physically handicapped person due to loss of a hand or a foot. One judgment committee even states that there is no chronic poisoning except for heavy metal poisoning. They cannot seem to understand alcoholism, glue sniffing, benzene poisoning, pesticide poisoning, drug addiction, and so forth. It becomes impossible for a patient suffering from CS to work in an ordinary office. Judgment committee members with common sense are needed for the protection of workers' lives.

3.2 UNDERSTANDING OF CS AT THE JUDICIAL LEVEL

Japanese judges are fundamentally very serious and excellent. They are tackling the problem of CS earnestly and the lives of many patients have been saved. Of course, the common sense of the time and the social situation form part of the legal decision. Naturally it is impossible to entrust all the CS patients' relief to a judge. If a judge gains further understanding of CS, a patient will experience greater relief. For a judge to understand the CS problem deeply, a medical staff's efforts are required.

3.3 EDUCATION

3.3.1 Specific Education of Medical Doctors

The classic concept of poisoning that doctors have been learning is important. However, it has not included that the consequences of acute intoxication and chronic intoxication are similar. As a result, many doctors do not fully understand chronic intoxication. For example, an understanding of chronic insecticide intoxication is still difficult for doctors at present. Moreover, the dose–response curve, which shows the relationship between the given dose of a chemical substance and the appearance of symptoms, has been commonly considered as polytonic rather than monotonic, that is, not one curve but many.

As an example, aspirin in the usual daily use amounts alleviates fever and has an analgesic action. However, a small dosage has also been shown to prevent thrombus formation. In the amount used for an analgesic action, a thrombus prevention action is not expected. Thus, the medicinal action differs from the usual dose to a small dose. Furthermore, a very small quantity of a chemical substance is enough to cause an endocrine disruption, aggravate an allergy, and induce CS. Also, the dose–response curve is not a sigmoid curve in classic poisoning studies but a bell shaped curve. If the chemical substance is much quantity, the action of hyper sensitivity is inhibited. The stimulating action much suppressed in high doses of chemicals.

A very small quantity of a chemical substance also causes inborn errors of metabolism, developmental disease, and behavioral abnormalities. It has been clearly proved experimentally that a very small quantity of an organophosphorus insecticide causes a developmental disorder. The doctors engaged in clinical work are well aware of the latest abnormal increase of a developmental disorder [1]. A very small quantity of a chemical substance can cause various kinds of diseases. Moreover, a single chemical substance has many actions. Invasion of chemical substances into everyday life is a concern. Education regarding chemical substances in the environment is required for doctors.

3.3.2 Specific Education for Dentists

Many patients react to a curative medicine. A dental care agent contains a small quantity, resulting in a sensitive reaction. General anesthesia is a kind of acute intoxication. Therefore, in general anesthesia, a sensitive reaction does not appear easily. Anesthesia using xenon should be considered for patients who can never receive general anesthesia.

Consideration of allergy to metal is important. Although dentists request patch tests from dermatologists during dental care using metals, it is possible for them to carry these out on their own. To test for a hyperreaction to cement and metal, a bit of material may be placed for 10 minutes under a patient's tongue and the patient observed for half a day to one day. To avoid a mechanical stimulus to the mucous membrane, the test materials are soaked in water before the test.

A patient sometimes shows a sensitive reaction to a topical anesthetic. It can be difficult to determine whether the chemical substance that causes the reaction is a component of the medicine itself or an additive.

At the time a sensitive reaction occurs, oxygen inhalation is the most effective treatment. It is common for the oxygen tolerance of CS patients to be low. Although extensive inhalation is needed at the time of anaphylaxis, 1–2 liters per minute of oxygen inhalation are usually recommended as soon as a reaction occurs. After consultations with a patient regarding a painkiller and an antibiotic according to a particular situation, the primary healthcare provider refer to the doctor or the dentist.

The indoor air of a dental clinic is usually polluted. Shortening a patient's length of stay in the waiting room as much as possible is required.

3.3.3 Specific Education for the Healthcare Team

Present-day medicine is team medical treatment. The goals of treatment of CS cannot be achieved only by a doctor's efforts. Cooperative efforts with a pharmacist, a nurse, an inspecting engineer, a dietitian, and an oral hygienist are important.

A patient with CS also often shows a sensitive reaction to a medicine. A patient may react to the active ingredient of a medicine, or to an additive. Pharmacists need to pay attention to patients' complaints just as doctors do. There is a tendency for a CS patient's detoxification function to be slower than that of an unaffected person. It is better to begin administering the medication in a small quantity except in an emergency.

When the patient with CS reacts to ethyl alcohol, chlorhexidine gluconate or benzalkonium chloride should be used. The patient often has a food allergy. Food additives and drinking water should also be considered. It is better for an inpatient to be hospitalized in a single room where ventilation by opening a window is possible. Nutritional support to the patient is important.

3.3.4 Patient Education for a Comfortable Daily Life

The present environment is filled with mental, physical, and chemical stressors. CS patients need to lower the total amount of environmental stressors.

3.3.5 Education of Laypersons

It is our duty to teach laypersons about current environmental pollution. Increases in inborn errors of metabolism, developmental disorders, anomalous behavior, and disorders of the immune system are increasing. The environmental pollutant which is dangerous for a CS patient is dangerous for a laypeople [1,2]. The power of laypersons in instigating change in company and government policies is very important.

ENDNOTES

1. Seinhausen HC, Dopfner M, Schubert I. 2016. Time trends in the frequencies of ADHD and stimulant medication. *Zeitschrift fur Kinder-und Jugendphychiatrie und Psychotherapie* 44:275–84.
2. Saadeh D, Salameh D, Caillaud D et al. 2015. Prevalence and association of asthma and allergic sensitization with dietary factors in schoolchildren data from the French six cities study. *BMC Public Health* 15:993.

4 Medical Facts

Mikio Miyata, PhD, MD, Professor Emeritus
Kitasato University
Kou Sakabe, PhD, MD, Dean and Professor
Medical School, Tokai University
Satoshi Ishikawa, PhD, MD, Professor Emeritus
Kitasato University

4.1 ACUTE TOXICOLOGY, INTERMEDIATE SYNDROME

A dose–response curve for acute intoxication is shown in Figure 4.1. If the amount of exposure increases, the condition will worsen and can culminate in death by intoxication, that is, the curve will take an S shape. Chronic intoxication, or the response to a very small quantity of a chemical substance.

4.2 CHRONIC TOXICOLOGY

Some decades ago, chronic intoxication was considered to be intoxication caused by chronic chemical exposure. However, the residual effect of acute intoxication is also increasingly considered to be equivalent to chronic intoxication. Central nervous system damage, peripheral neuropathy, disturbances in immune function, endocrine disruption, and so forth are included as manifestations of chronic intoxication. An example of chronic intoxication, by an organophosphorus (OP) pesticide, is described in Chapter 1. Developmental disease, inborn errors of metabolism, and abnormalities in immune responses have now been added to the previously identified risks.

A very small quantity of a chemical substance aggravates allergies, chemical sensitivity (CS), abnormal central nervous system reactions, and disruptions of the endocrine system. These are overreactions to a chemical substance, which are different from poisoning.

4.3 ANIMAL EXPERIMENTS USING EXTREMELY LOW DOSAGES

CS experiments in animals are difficult. But when one considers how a chemical substance acts on a living being, it is necessary to remember "the law of Arndt-Schulz," a fundamental concept of pharmacology. It states that "a little of a chemical substance stimulates a function although a lot of a chemical substance suppresses a function." Suppression by a large amount of a chemical substance means disruptions in the function of the organs and possibly death by poisoning. Further, the chemical substance may have many actions. For example, vitamin C has more than 40 kinds of actions.

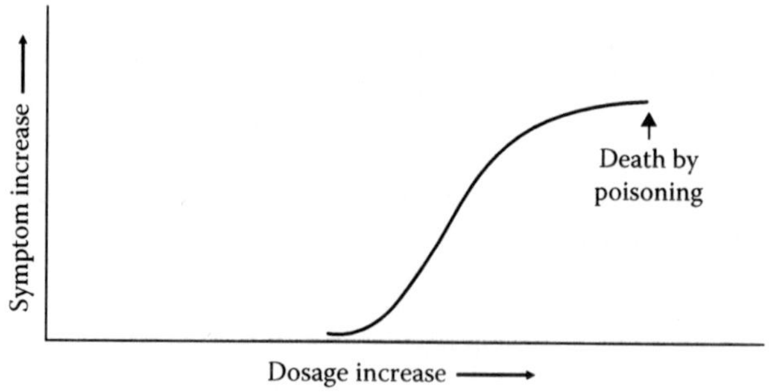

FIGURE 4.1 Dose–response curve.

In experiments in which rats were poisoned with fenthion (an OP), a stimulus action was shown with the small quantity of 1/40,000 of LD_{50} [1].

Allergy is a hypersensitive reaction of in immune system. CS is a hypersensitive reaction centering on a nervous system. The nervous system are closely in cooperation with the immune system. The patient with CS often suffer from allergic disease. Allergy is influence trace chemicals.. Experiments on hay fever clearly show that a very small quantity of a chemical substance triggers an allergic reaction. As an example, effect of a water-soluble OP, trichlorfon, in experimental allergic conjunctivitis is shown in (Figure 4.2). The allergic reaction was intensified by a very small quantity, 30 ng/kg by weight [2].

However, when the quantity was increased further, the allergic reaction decreased. As a result, the reaction shows a bell-shaped curve. Until now, we could draw a

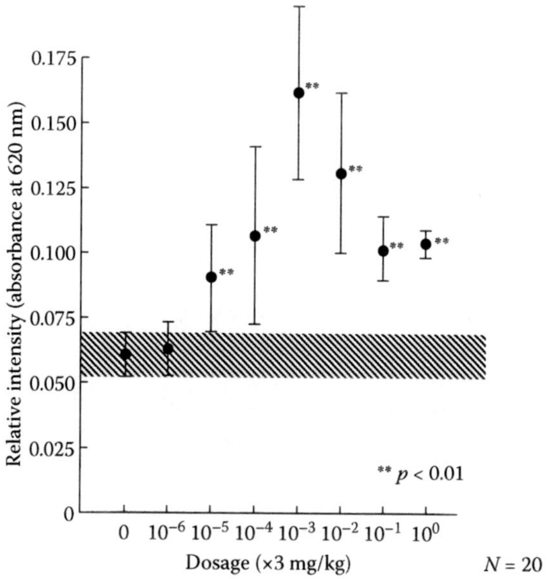

FIGURE 4.2 Reaction curve of trichlorfon for experimental allergic conjunctivitis.

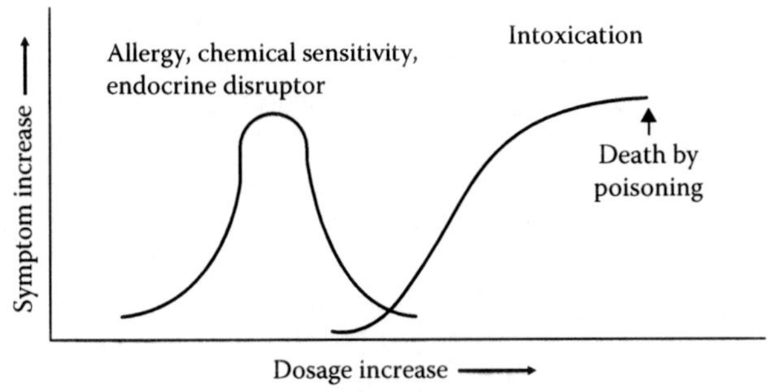

FIGURE 4.3 Graph showing the overlap of a classic dose–response curve in intoxication as sigmoid curve and in hyperactivation curve for trace chemicals as bell shaped curve.

similar reaction curve for weed killers, trihalomethane, paradichlorobenzene [3], synthetic food colorants, refined sugar, and so forth. A bell-shaped curve has also been observed for endocrine disruptors.

If this relation curve is overlapped with the dose–response curve of a previous intoxication, a key map (Figure 4.3) can be shown.

It is understood that it is common for a CS patient to react to a very small quantity of a multitype chemical substance.

4.4 RESEARCH ON THE MECHANISM OF CS

Why does CS not affect all only affect some members of a population. The influence of environmental chemicals on individuals is determined by several factors: (1) hereditary and individual biochemical differences, (2) whole body loading dosage, (3) age and sex, (4) underlying disease, (5) nutrition, and so forth. Differences in these factors account for the "individuality" that is one of fundamental concept of CS. In research on the action of a chemical (xenobiotic) substance in the body involving disruption of endocrine function, the action of bisphenol A did not follow the pattern of a monotonic dose–response curve seen in classic poisoning. A very small quantity of a chemical substance shows a completely different, polytonic, dose–response curve [4].

It has been commonly recognized that the symptoms of CS develop from chemical substance exposure in many cases, and a nervous system sensitivity reaction occurs. In this section, we consider the mechanisms of detoxification of a chemical substance and production of a sensitivity reaction in CS.

4.4.1 DETOXIFICATION

Decomposition of a xenobiotic substance in the human body is roughly divided into two reaction steps. In the first step, a chemical substance undergoes a reaction such as oxidization, reduction, a hydrolysis reaction. The metabolic product of this first enzymatic reaction creates oxidative stress. Therefore, it is quickly transferred to the enzyme reaction of the second step, which is a conjugation reaction with glutathione, glucuronic acid, sulfuric acid, amino acids. Many lipophilic chemical substances thereby acquire water solubility and can be easily released from the body.

P-450 (CYP) is the main group of enzymes involved in the first step of the decomposition of a chemical substance. Cytochrome P-450 is an enzyme group widely present in the body. Although humans have been exposed to countless chemical substances since the Industrial Revolution, they have usually responded reasonably well because the CYP family, consisting of CYP1, CYP2, CYP3, and CYP4, metabolized various environmental chemicals. Genetic polymorphism is observed in these CYPs.

Prang has reported that high levels of CYP1A2 and 2D6 are present in patients with CS. This research provides important insights in showing that the intermediate metabolite is superfluous in expression of CYP1A2 and 2D6 [5]. These patients are considered to be in a superfluous oxidative stress state.

Neurotoxic esterase (NTE) is another important enzyme to be considered in relation to OP compounds. More than half of CS cases are said to be caused by OP. Mice with low NTE activity showed hyperactivity caused by a small amount of OP as compared with normal mice [6]. The systemic disturbance of the nervous system of patients with CS resembles the Gulf War syndrome among soldiers fighting in the Persian Gulf War that broke out in 1990. Soldiers with nervous system conditions have been suspected to have a history of exposure to OP pesticides. NTE activity was found to be reduced in these soldiers and was attributed to contact with nerve gas [7–10]. Individual genetic differences have also been proven in paraoxonase (PON) regarding detoxification of OP [11,12].

Many enzymes such as glutathione-*S*-transferase, glutathione peroxidase, gluculonosyl transferase and *N*-acetyltransferase are included in the second step.

Glutathione-*S*-transferase (GST) catalyzes glutathione conjugation. Prang also reported the gene polymorphism of GST in patients with CS and posited that this enzyme has a low heritability or a slow reaction velocity [5]. The deficit of the gene *GSTM1* in some patients was proven in our 15 patients [13]. All patients were very carefully selected and diagnosed with CS at Kitasato University Hospital (Figure 4.4).

De Luca [14] reported no difference in genotype of detoxification, such as glucu-lonosyl transferase, and cytochrome P-450. However, in patients with CS, levels of catalase in red corpuscles and glutathione *S*-transferase were low while glutathione peroxidase levels were high. Oxidized and reduced glutathione levels decreased and

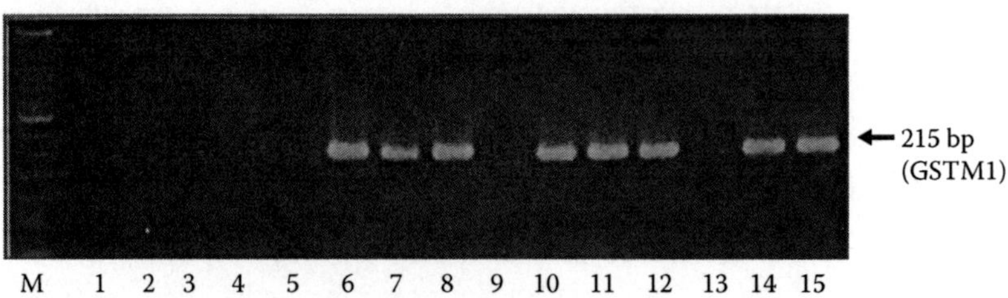

M: bp marker. Case 1, 2, 3, 4, 5, 9, 13: defect of GSTM1.

FIGURE 4.4 Glutathione *S*-transferase in patients with CS.

levels of nitric acid and a nitrous acid compound increased. Fatty acids tended to be saturated fatty acid. Inflammatory cytokines were increased. These results are well in agreement with clinical experiences.

4.4.2 Nervous Sensitivity Acquisition

Communication occurs between the nervous, endocrine, and immune systems (Figure 4.5).

The limbic system, located under the cerebrum, consists of several different cortical and subcortical structures and supports numerous functions essential to life, including emotional responses and motivation. The limbic system influences the endocrine system, including the hypothalamus–pituitary–adrenal axis, which is the center of the stress reaction. Many diseases can affect this system, such as chemical sensitivity, fibromyalgia, chronic fatigue syndrome, organic phosphorus intoxication, Gulf War syndrome, allergy, and organic chlorine intoxication. Environmental pollution has also been considered to contribute to anaphylactic enteritis syndrome and developmental disorder.

To date, the mechanism of CS has been discussed mainly in the context of effects of chemical substances on the limbic system and their involvement in the kindling phenomenon [15].

In the limbic system, external sensory information is conveyed through two routes: the frontal association area and the hippocampus through the Papez circuit and the amygdala through the Yakovlev circuit. The hippocampus is a center of memory and emotion and the amygdala is the center of smell. The Papez and Yakovlev circuits interact. The hippocampus is considered to be a focal point for CS.

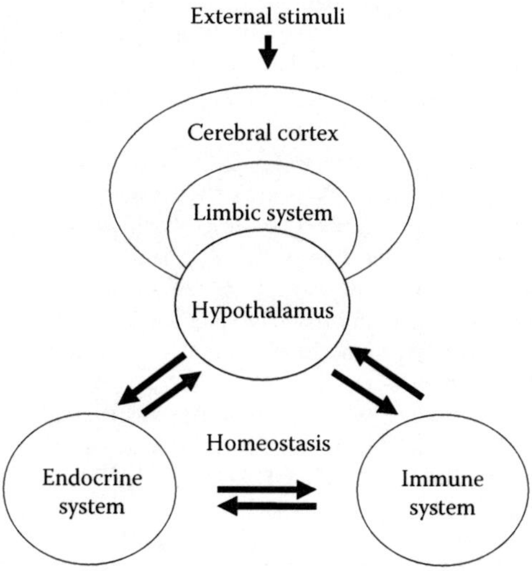

FIGURE 4.5 The nervous, endocrine, and immune systems comprise the functional axis of homeostasis.

In the kindling, phenomenon, a nerve begins to react to a weak stimulus when it is repeated, which would not occur in response to a single stimulus [16].

The stimulus reinforcement is present in a peripheral sensory nerve. If the unmyelinated nerve fiber (C fiber) end of the sensory nerve in the respiratory tract membrane and elsewhere is stimulated by a chemical substance, the stimulus will be conducted to a center. Simultaneously, after the stimulus changes the direction of transmission in another branch of the neuron, it is transmitted convertedly to a tip and tachykinin is secreted. Substance P, among other neuropeptides, is a member of the tachykinin family. It induces the release of histamine from mast cells and it causes activation of immune cells, such as lymphocytes, macrophages, neutrophils, and eosinophils. Moreover, tachykinin causes an increase in the branches of the nerve end of C fibers. Thus, a chemical substance causes an increase in neurogenic inflammation, creating a vicious cycle. If a chemical substance repeatedly stimulates through olfactory mucosa and the olfactory nerve, a kindling phenomenon can be induced in the limbic system that is thought to have consequences for mental state and nervous system function. These phenomena have been demonstrated in animal experiments [17,18].

ENDNOTES

1. Miyata M, Imai H, Ishikawa S. 1973. Electroretinographic study of the rat after fenthion intoxication. *Jpn J Ophthalmol* 17:335–343.
2. Namba T, Honma K, Horiuchi, K, Tujisawa I, Miyata M, Ishikawa S. 1993. Participation of environmental chemicals in experimental allergic conjunctivitis. *Acta Soc Ophthalmol Jpn* 97:297–303.
3. Li G, Hanai Y, Miyata M, Ishikawa S. 1994. Aggravating effects of chloroform and *p*-dichlorobenzene on experimental allergic conjunctivitis. *Folia Ophthalmol Jpn* 45:475–480.
4. Jenkins S, Wang J, Eltoum I, Desmond R. 2011. Chronic oral exposure to bisphenol A results in a nonmonotoic dose response in mammary carcinogenesis and metastasis in MMTV-erbB2 mice. *Environ Health Perspect* 119:1604–1609.
5. Prang NS, von Baehr V, Bieger WP. 2001. Genetische Susceptibilitaet gegenueber Umbeltgiften. *Z Umweltmedizine* 9:38.
6. Winrow CJ, Hemming ML, Allen DM. 2003. Loss of neuropathy target esterase in mice links organophosphosphate exposure to hyperactivity. *Nat Genet* 33:477–485.
7. Ashford N, Miller C. 1998. *Gulf War exposure, chemical exposure, low levels and high stakes*, pp. 246–249. New York: John Wiley & Sons.
8. Mutch E, Blain PG, Williams FM. 1992. Interindividual variations in enzymes controlling organophosphate toxicity in man. *Hum Exp Toxicol* 11:109–116.
9. Colomba BA. 2008. Acetylcholinesterase inhibitors and Gulf War illnesses. *Prog Nati Acad Sci USA* 105:4296–300.
10. Casida JE, Quistad GB. 2004. Organophosphate toxicology: Safety aspects of nonacetylcholinesterase secondary targets. *Chem Res Toxicol* 17:983–998.
11. McKeown-Eyssen G, Baines C, Cole DE et al. 2004. Case control study of genotypes in multiple chemical sensitivity: CYP2D6, NAT1, NAT2, PON1, PON 2 and MTHFR. *Int J Epidemiol* 33:971–978.
12. Costa LG, Cole TB, Furlong CE. 2003. Polymorphisms of paraoxionase (PON1) and their significance in clinical toxicology of organophosphates. *J Toxicol Clin Toxicol* 41:37–45.

13. Sakabe K, Miyata M, Ishikawa S. 2003. Thinking on chemical sensitivity. *Neuro-ophthalmol Jpn* 20:350–354.
14. De Luca C, Scordo MG, Cesareo E et al. 2010. Biological definition of multiple chemical sensitivity from redox state and cytokine profiling and not from polymorphisms of xenobiotic-metabolizing enzymes. *Toxicol Appl Pharmacol* 248:285–292.
15. Tran MT, Arendt-Nielsen L, Kupers R et al. 2013. Multiple chemical sensitivity: On the scent of central sensitization. *Int J Hyg Environ Health* 216:202–210.
16. Gilbert ME. 2001. Does the kindling model of epilepsy contribute to our understanding of multiple chemical sensitivity. *Ann NY Acad Sci* 933:68–91.
17. Adamec R, Young B. 2000. Neuroplasticity in specific limbic system circuits may mediate specific kindling induced changes in animal affect-implications for understanding anxiety association with epilepsy. *Neurosci Biobehav Rev* 24:705–723.
18. Botterill JJ, Foumier NM, Guskjolen AJ et al. 2014. Amygdaia kindling disrupts trace and delay fear conditioning with parallel changes in Fos protein. *Neuroscience* 265:158–171.

5 Diagnosis

Satoshi Ishikawa, PhD, MD, Professor Emeritus
Kitasato University
Mikio Miyata, PhD, MD, Professor Emeritus
Kitasato University

5.1 DETAILED AND CAREFUL INTERVIEW

The first step in the diagnosis of sick-house syndrome (SHS) and chemical sensitivity (CS) begins with determining the etiological factor. Before examining a patient, doctors in the Kitasato University clinic send an interview sheet including 11 pages of the Quick Environmental Exposure and Sensitivity Inventory (QEESI®) by C. S. Miller [1]. The questionnaire includes the history of local and systematic allergy, medical treatment in dental procedures, occupation, birthplace, relocation and reconstruction/modification, living environment, the medicines used, food, luxury goods, and so forth. Ishikawa partly modified this questionnaire into the Japanese language and examined CS cases. Those studies showed similar results in the symptoms between US and Japanese patients [2].

Although Randolph identified CS in the 1960s, his statements were disregarded [3,4].

In the United States, a CS case was diagnosed in 1999, according to six agreed on factors in the diagnosis of CS [5]:

1. "The symptoms are reproducible with repeated chemical exposure."
2. "The condition is chronic."
3. "Low-level exposure lower than previously or commonly tolerated results in manifestations of the syndrome."
4. "When the incitants are removed, the symptoms improve or resolve."
5. "Response occurs to multiple chemically unrelated substances."
6. [Added in 1999]: "Symptoms involve multiple organ systems."

In Japan, neuro-ophthalmological examinations are included with ordinary outpatient clinical examinations in the diagnostic criteria of CS (Table 5.1) [6].

5.2 NEURO-OPHTHALMOLOGICAL AND NEUROLOGICAL EXAMINATIONS IN CONJUNCTION WITH ORDINARY CLINICAL EXAMINATIONS IN JAPAN

Some of the following neurological tests are the diagnostic procedures that the author developed in the 1980s. Parts of those examination methods have been used mostly in the United States by several pioneers: as K. H. Kilburn [7].

TABLE 5.1
Diagnostic Criteria for Chemical Sensitivity in Japan

A. Cardinal symptoms: Headache, muscular pain, fatigue, languor
B. Subsymptoms: Throat pain, slight fever, diarrhea, constipation, dazzle, decline in thinking ability and concentration, emotional instability, excitement, insomnia, forgetfulness, itchiness, the abnormalities in feeling, irregular menstruation
C. Positive laboratory findings: Pupillary reaction, decline of visual sensorial sensitivity, abnormalities in eye movement, and functional deterioration of the cerebral cortex

5.2.1 Tracking Eye Movement Examination

A tracking eye movement examination including smooth pursuit and saccadic eye movement is performed to determine dysfunctions of eye movement. In CS patients, defective smooth pursuit eye movement is frequently seen; as one example, such an abnormality is often seen in chronic intoxication with an organophosphorus (OP) insecticide. A skilled doctor can judge smooth pursuit movement with the naked eye.

Figure 5.1 shows an example of smooth pursuit movement in a healthy person (left) and in a patient with CS (right). The motion of the CS patient's eyes is not smooth like that of an unaffected person's eyes. Eye movement is damaged in approximately 70% to 80% of CS patients.

Smooth pursuit movement of a 21-year-old female patient with CS who improved with medical treatment is shown in Figure 5.2. Smooth pursuit defects before therapy (left, with marked staircase trajectories) have been well recovered after therapy (right, with normal pursuit).

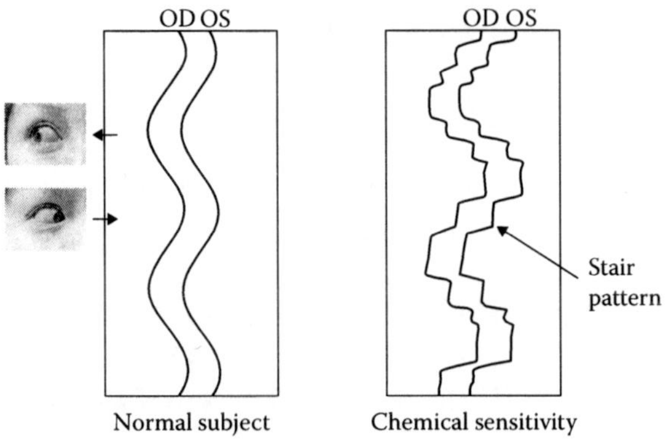

FIGURE 5.1 Smooth pursuit eye movement in a normal person and in a patient with CS. The eyes move horizontally to the OD (right eye) and OS (left eye). The movement of both eyes is recorded.

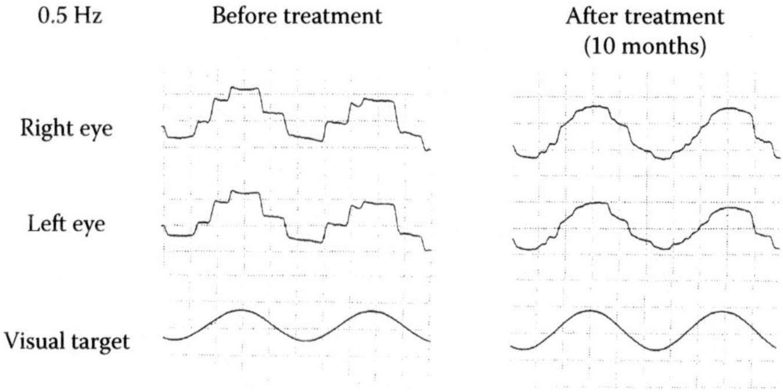

FIGURE 5.2 Smooth pursuit vertical eye movements of 0.5-Hz loading improved 10 months after the treatment of CS. Abnormal staircase pursuit of the eye movements was markedly improved.

5.2.2 Pupillary Function

CS patients experience both central and autonomic nervous system dysfunction. This abnormality of automatic nervous system dysfunction can be well diagnosed by examining the pupillary reaction to light stimuli. A patient goes into a darkroom to let his or her pupil dilate by pre-dark adaptation. About 10 minutes of pre-dark adaptation is necessary to let the pupil fully dilate. The dilated pupil constricts when light enters the pupil. After the light stimulus is turned off, the pupils dilate once again in a dark room. The constriction reaction is controlled by the parasympathetic nerve and the dilatation is controlled by the sympathetic nervous system. The change in the pupil is shown in Figure 5.3.

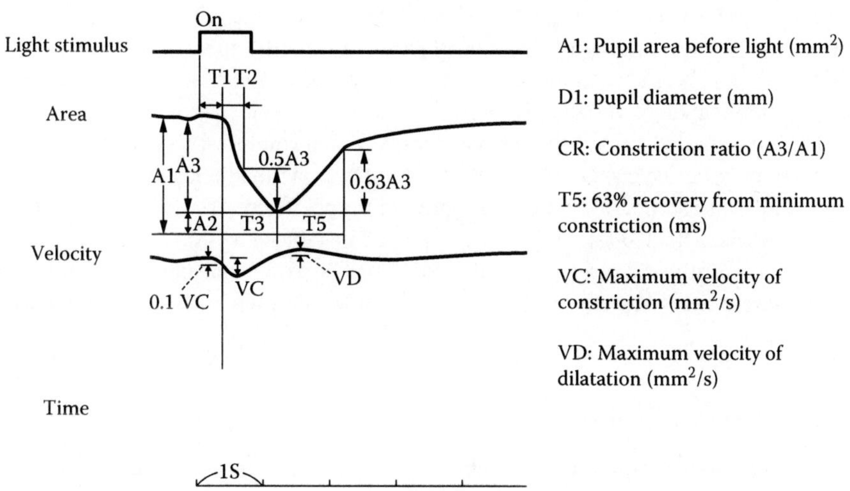

FIGURE 5.3 Schematic drawing of light reaction: Pupil response and various variables during constriction and dilatation.

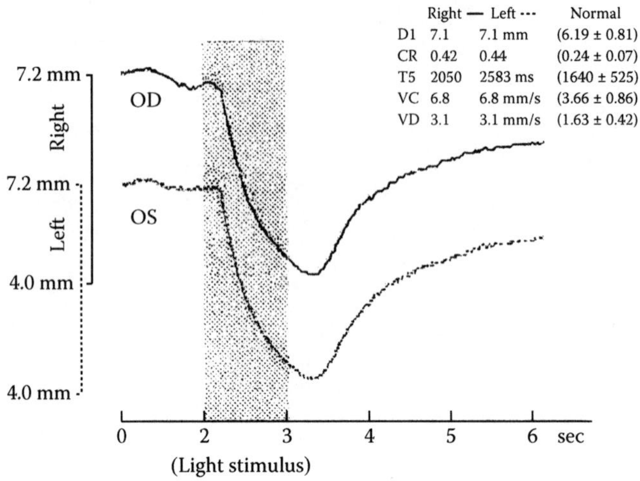

FIGURE 5.4 Actual trajectories of pupillary responses in a patient with CS. OD (right eye) and OS (left eye).

Frequently used clinical assessments include the pupil horizontal diameter; D1, constriction rate; VC, velocity of constriction; VD, velocity dilatation; and T5, the time that the constricted pupil returns to 63% of the original size.

Actual recordings obtained from a 37-year-old man who showed symptoms of CS from indoor air pollution in the company where he was employed are shown in Figure 5.4. The right (upper pupillogram) and the left eye (the lower pupillogram) are shown. The hatched area is the duration of the light stimulus. Each measured value is indicated in the upper right frame. The value of measurements, even in a normal pupil reaction, differ according to sex and age and therefore must be judged against the normal values obtained from the subjects. In this patient's record, the CR, T5, VC, and VD values exceed the range of normal values. A cholinergic type of disorder was confirmed.

The examination can reveal various pupillary abnormalities: a sympathetic nerve strain type, a parasympathetic nerve strain type, a sympathetic nerve suppression type, and a parasympathetic nerve suppression type. There are also complex types that are unclassifiable and pupil unrest type, ie., instable type [8].

Autonomic imbalance is detected in 65% of CS patients at pupil examinations, which have a high objectivity. Recovery of autonomic imbalance in CS patients is slow.

5.2.3 Standing Ability

We use the center-of-gravity fluctuation meter as one of the tests of body balance ability. A patient stands on a center-of-gravity meter for 1 minute with eyes open followed by 1 minute with eyes close. The standing ability of patients with CS is often disturbed [9].

5.2.4 Contrast Sensitivity Examination of Higher Visual Centers

Visual contrast sensitivity is a high-order function of cerebral vision, different from visual acuity. A lattice pattern of black and white in which boundary between black and white is indistinct is used in an actual examination.

Even CS patients with good eyesight frequently say that it is hard to see. A decline in contrast sensitivity is detected in 85% of CS patients [10].

5.2.5 Accommodation Examination of Eyes

The accommodative function can decline in CS patients. A visual object moves from far to near after moving from near to far. Ocular accommodation is then recorded. Some type of accommodation disorder is detectable in 65% of CS patients [11]. The drawback of this method is that it is not useful in older patients.

5.2.6 Cerebral Function Examination

To confirm functional brain disorders of CS patients, single-photon emission computed tomography (SPECT), positron emission tomography (PET), functional magnetic resonance imaging (fMRI), and other imaging modalities have been used and disorders of cerebral function and blood flow obstruction have been reported [12,13]. Near-infrared spectroscopy (NIRS) is frequently used in Japan.

In NIRS, laser light of the near-infrared wavelength is irradiated at the frontal region of the brain. Most of the light is absorbed, such as in oxygenated hemoglobin. A small amount of light will be emitted from the head as a reflection. NIRS measures the strength of the weak light that was reflected (Figure 5.5).

We measured the change in blood flow in the brains of patients with chemical sensitivity by single-channel NIRS. A tendency toward change in cerebral blood flow change was observed [9]. A recent Japanese study using a multichannel NIRS reported that following exposure to odorous chemicals, patients with CS showed significantly increased blood flow in the prefrontal cortex compared to the controls. The prefrontal cortex plays a key role in many sensory and motor functions.

The validity of these methods, such as finding hot and cold foci in a brain through SPECT, has been reported [14]. Using SPECT, we have also found abnormalities in 58% of CS patients [15].

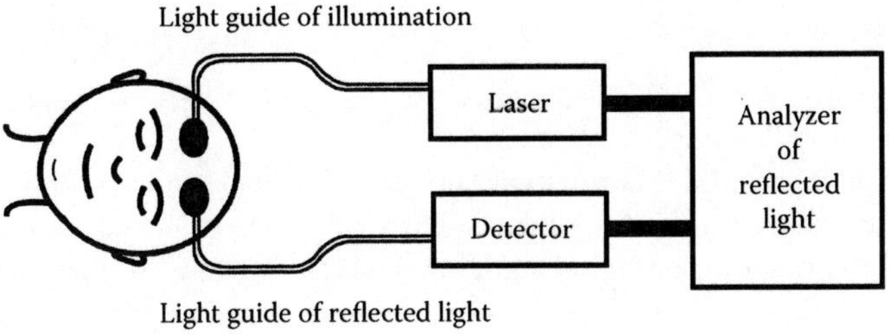

FIGURE 5.5 Near-infrared spectroscopy (NIRS).

5.2.7 Immune Examination

The nervous system and the immune system have a inter relationship. Recently, a significantly elevated pro-inflammatory cytokine, interleukin-1-beta, -2, -4, and -6, in the plasma of CS patients was reported. Tumor necrosis factor-alpha was significantly increased. Interleukin-13 was significantly decreased [14]. A future supplementary examination is required. It is expected that abnormal findings in the immune system will help in diagnosis as well as abnormal findings in the nervous system.

5.2.8 Respiratory Function

Few CS patients show abnormalities of a fall of V50 and V25 [16]. Bascom also observed problems of respiratory function in CS patients' [17].

5.2.9 Peripheral Venous Blood Oxygen Concentration

A high partial pressure of oxygen in peripheral venous blood may be seen in patients with CS, as noted by W. Rea [18]. In his view, the ability to use oxygen, accompanying a respiratory functional improvement of a cell, in CS patients is improved by oxygen inhalation. This concept is completely correct in clinical practice.

5.3 CHEMICAL LOAD TESTS

5.3.1 Use of a Clean Room

For a chemical substance load test, as in the antigen load examination of allergy, removal of masking is desired. It means removal of the acclimation to an environmental pollution of a chemical substance. The patient needs to stay in a stable clean space, without chemical pollution, during a certain period for removal of the acclimation state. In the clean room, the floor, walls, ceiling, thermal insulation, fixtures, and so forth are so constructed as to make emission of an air pollution substance difficult. An air conditioner that sends pure air indoors is required.

A HEPA (high-efficiency particulate arrestance) filter and activated charcoal filter supply clean air indoors. The entrance and inside of an actual clean room are shown in Figure 5.6.

5.3.2 Determination of Causative Chemicals (Challenge Test)

A challenge test is carried out quantitatively. It is necessary to prepare a special separate room for loading.

A formaldehyde load examination supported by the Ministry of Environment Japan was conducted in 37 CS patients. The load of formaldehyde gas was 40 and 8 ppb, and the room air pollution guideline value in Japan is 80 ppb. Although the load trial was performed in a double-blind manner, various problems emerged. It became clear through both formaldehyde and placebo load that the patients were in excessive strain states. Maximum blood pressure fell at the end of the load test when

FIGURE 5.6 Entrance of a clean room for CS patients with positive air pressure control to prevent entry of chemicals pollution from the outside. The room is also shielded against electromagnetic waves.

both formaldehyde and a placebo were used. The extent of recovery of maximum blood pressure after formaldehyde was larger than after placebo. It means that the stress caused by formaldehyde is larger than that caused by placebo. Changes of the subjective symptoms in these 38 persons are shown in Table 5.2.

As an objective study, the pupillary reaction test for evaluation of the autonomic nervous system (Table 5.3) were examined. Refer to Figure 5.3 for the indicated variables. A number of cases in the table showed a significant change in the variables.

TABLE 5.2
Subjective Reaction to Load Test

Correct Reaction	Incorrect Reaction	Difficult to Judge
21 cases	7 cases	10 cases

TABLE 5.3
Variables Change after Challenge Test (N = 37)

Placebo						8 ppb						40 ppb					
A1	A3	CR	T5	VC	VD	A1	A3	CR	T5	VC	VD	A1	A3	CR	T5	VC	VD
6	5	6	0	1	0	11	10	8	7	9	6	8	8	9	8	9	6

Formaldehyde load markedly affects the autonomic nervous system. A significant difference was seen in T5 between placebo and the 8 and 40 ppb loadings. There was no significant difference between placebo and formaldehyde in other parameters [10].

A chemical substance load examination is an effective method. But even if a reaction to one chemical substance is proven, it is difficult to determine that the substance is the real cause of the onset of illness. Moreover, symptoms become more severe in some patients after a load examination.

ENDNOTES

1. Miller CS, Prihoda TJ. 1999. The environmental exposure and sensitivity inventory (EESI). A standardized approach for measuring chemical intolerances for research and clinical application. *Toxicol Indust Health* 15:370–385.
2. Hojo S, Yoshino H, Kumano H, Kakuta K, Miyata M, Ishikawa S et al. 2005. Use of QEESI Questionaire for screening study in Japan. *Toxicol Indust Health* 21:113–124.
3. Randolph TG. 1961 Human ecology and susceptibility to the chemical environment, parts I and II. *Ann Allergy* 19:516–540.
4. Randolph TG. 1964 The ecologic unit. Part 1. *Hosp Manage* 97:45–47.
5. Multiple chemical sensitivity: A 1999 consensus. 1999. *Arch Environ Health* 54:147–149.
6. Ishikawa S. 1998. Chemical sensitivity; diagnostic criteria. *Jpn Med J* 3857:25–29.
7. Kilburn, KH. 1998. *Chemical brain injury*, pp. 6–91. New York: Van Nostrand Reinhold.
8. Ministry of the Environment Japan, Japan Public Health Association. 2001. Research report of idiopathic multiple chemical sensitivity for the 2000 financial year.
9. Research report of Multiple chemical sensitivity in 2000 financial year. 2001. Japan Public Health Association sponsored by Ministry of Environment Japan.
10. Miyata M, Namba T. 1996. Clinical aspects of multiple chemical sensitivity. *Auton Nerv Syst* 33:257–261.
11. Kikuchi H, Ichibe Y, Namba T, Miyata M, Ishikawa S. 2000. Neurological-ophthalmological findings of patients with chemical sensitivity. *Jpn J Clin Ecol* 9:22–27.
12. Andersson L, Claesson AS, Nyberg L et al. 2014. Brain responses to olfactory and trigeminal exposure in idiopathic environmental illness (IEI) attributed to smells-an fMRI study. *J Psychosomatic Res* 77:401–408.
13. Hillert L, Jovanovic H, Ash F et al. 2013. Women with multiple chemical sensitivity have increased harm avoidance and reduced 5-HT (IA) receptor binding potential in the anterior cingulate and amygdala. *PLOS ONE [Electric Resonance]* 8: e54781.
14. Dantoft TM, Elberling J, Brix S, Szecsi PB, Vesterhauge S, Skovbjerg S. 2014. An elevated pro-inflammatory cytokine profile in multiple chemical sensitivity. *Psychoneuroendocrinology* 40:140–150.
15. Simon TR. 1991. Single photon emission computed tomography of the brain in patients with chemical sensitivities. *Toxicol Indust Health* 10:542.
16. Ministry of the Environment Japan, Japan Public Health Association. 2003. Research report of idiopathic multiple chemical sensitivity for the 2002 financial year.
17. Bascom R. 1992. Multiple chemical sensitivity: A respiratory disorder. *Toxicol Indust Health* 8:221–228.
18. Rea WJ. 1992. *Chemical sensitivity*, Vol. 4, pp. 2554–2563, Ririe, ID: Lewis Publishing.

6 Treatment of Patients with Chemical Sensitivity

Mikio Miyata, PhD, MD, Professor Emeritus
Kitasato University
Satoshi Ishikawa, PhD, MD, Professor Emeritus
Kitasato University

Many medical treatment methods for chemical sensitivity (CS) have been suggested that have not yet been sufficiently proven scientifically. Clinicians promote treatments based on their own preferences and perceptions, and there are some differences among them. The fundamental principles governing medical treatment are first to decrease the amount of the chemical (xenobiotic) substance in the body; second, to eliminate the adverse reactions that xenobiotic substances cause; and third, to ensure mental stability.

6.1 AVOIDANCE OF CHEMICAL SUBSTANCES

It is most important to prevent inflow of the chemical substance to the body. Moreover, simultaneous reduction of the total stressors, including physical (temperature, sound, vibration, light, electromagnetic waves) and biological stressors is required.

6.1.1 Food

Food additives are pervasive. The quantity of food additive consumed by one person every day is said to be not less than 20 g [1]. In present-day life, it is impossible to remove an artificial food additive completely.

6.1.2 Water

A living organism is an electric phenomenon in water. Contamination of the base element of life is getting worse. The trihalomethanes generated by chlorination of water worsen not only CS but also allergies. If person with atopic dermatitis and one with asthma swim in the pool of a hotel, their conditions will worsen within an hour. Human skin easily passes a lipid-soluble chemical substance to the interior of the body. Guinea pig experiments proved that a very small quantity of trihalomethane worsens allergies [2].

For typical CS patients, a water purifier is indispensable. Some CS patients, however, react to the substances in the water from a water purifier, and some CS patients cannot drink bottled water, apparently reacting to the elution substance from a cap, and so on. To avoid chemical substances present in water, some people drink distilled water prepared using glass equipment.

6.1.3 Air

An adult inhales about 20 kg of air in one day. Chemical substances in food and drink are absorbed and passed through the liver, which acts as a barrier and a gateway for detoxification. Conversely, chemical substances inhaled from air will enter blood directly, barrier free. Moreover, the chemical substance adhering to the nasal cavity membrane traveling along the olfactory nerve and directly reaches the brain. Therefore, chemicals in air pollution cause serious damage to the body and are a major contributor to CS symptoms.

Many chemical substances in air pollution have a specific gravity higher than that of air. Therefore, they collect easily at lower heights. It is thus easy for a child to inhale a chemical substance present in air pollution. Moreover, there are many air intakes per weight among children, and the detoxification capability of children is low. Further, many schools do not have air conditioning. According to Dr. Rapp of New York University, room air pollution of a school makes it a sick school [3]. Attention must be paid also to the volatile matter from the acoustic insulation of a music room, the vibration isolation material of the floor of a gymnasium, the equipment of a computer lab, the volatile matter of an art room, and so forth in a school.

A ventilation fan must always be used. The interior of a room will then be at negative pressure and the fan will be able to suck up a chemical contaminant from the bottom of a floor and the back side of a wall. In flow of air has to be kept simultaneously.

Volatile chemical substances from furniture, curtains, and floor wax are also extensive. Nitrogen oxide is contained in the exhaust gas from a kerosene stove. The air in a new car is filled with volatile chemical substances. The amount of chemical substances in air pollution increases sharply with a rise in temperature.

6.1.3.1 Use of an Air Purifier

There are three kinds of air purifiers: active charcoal, catalyst, and ozone. Each method has advantages and disadvantages.

The active charcoal system needs the activated carbon to be changed periodically and carries a maintenance cost. Some people have a reaction to the smell from active charcoal. Moreover, the activity of active charcoal is reduced in humid conditions.

The catalyst system has a low maintenance cost. However, there is no guarantee that a chemical contaminant is completely dissociated into water and carbon dioxide by the catalyst.

In an ozone oxidization system as well as a catalyst, the intermediary metabolite in the middle of decomposition may flow out. Moreover, the danger of ozone itself should also be considered.

If an air cleaner is used in the interior of a completely sealed room, the chemical substance in an air cleaner not intended for cleaning will accumulate.

Blowing clean air into the room is effective, as shown in Figure 6.1.

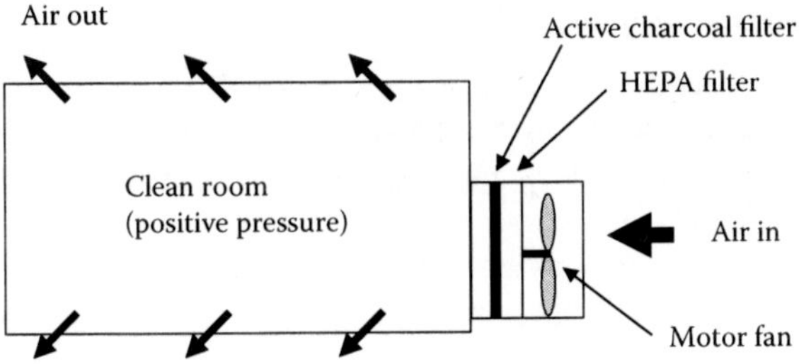

FIGURE 6.1 Blowing of clean air into a room.

The typical air cleaner is a recycled form of indoor air. This type of room is set as the interior of a room keeps a positive pressure. It can prevent inflow of outdoor contaminated air. The air contamination from motors can be absorbed also. However, the cost increases.

6.1.4 Clothes

Clothes often adsorb surrounding chemical substances from the air and can be a source of CS. Therefore, people with CS may not be able to wear clothes exposed to a polluted environment. Education regarding storage that does not use an insecticide and use of safe detergents is also required. Fundamentally, a natural fiber is the safest material for underwear. Moreover, a sensitive patient reacts also to the stitching of underwear. Turning underwear inside out is recommended in such situations.

6.1.5 Medicine

Seventy percent or more of CS patients have a history of an allergy [4]. Many patients also have drug allergies. Chemical substances including various diluent bases, antioxidants, antibacterial chemicals, artificial sweetening, artificial coloring materials, and so on are contained in the medicine in addition to its active ingredient. Even if a patient does not have an adverse reaction to the active ingredient, he or she may appear to react to a medicine in response to various additives. Moreover, although the debate continues, there are also some reports that the action of a patient's detoxification enzyme is slow. Medication dosage should be carefully monitored. Medication for a psychiatric illness may sometimes actually worsen the condition. The physician may then think that the dosage is insufficient and increase it, exacerbating the patient's condition. However, when a condition such as depression is serious, psychiatric medicines are used. Moreover, an antibiotic is sometimes indispensable after dental care and surgery. Medication should not be prescribed recklessly, but not necessarily denial.

6.2 DETOXIFICATION

Detoxification is of utmost importance in medical treatment.

6.2.1 Spa, Hot Spring, Bath, Low-Temperature Sauna

Thermotherapy is effective in detoxification. However, because many CS patients experience autonomic imbalance, hot bathing should be avoided. A slightly more tepid long bath is recommended. A hip bath and a foot bath are also effective in situations of autonomic imbalance.

A high-temperature sauna produces stress and therefore a low-temperature sauna is preferable. A bedrock bath, china plate bath, sand bath, and so on can likewise be used.

6.2.2 Drug Treatment

Chemical substances inside the body are called xenobiotics. These pass along various courses and are discharged to the exterior of the body in response to a structural change. The famous Hubbard Regimen is a method of treating xenobiotics.

The detoxification course is composed by two phases. The first phase includes oxidization, hydrolysis, and reduction. The second phase includes conjugation, acetylation, and methylation. Although detoxification generally progresses to phase II from phase I, there is also a course in which this does not occur. The action of glutathione *S*-transferase and *N*-acetyltransferase in CS patients is reported to be hereditarily slow, but the debate continues. An effective phase II process is the foundation of detoxification.

6.2.2.1 Conjugation

Glucuronate: Glucuronic acid tablets are used to promote glucuronic acid conjugation. Usually, 1 g of glucuronic acid is taken after meals three times a day, for a total daily dose of 3 g.

Glutathione: Although the human body produces glutathione naturally, it is prescribed for patients to promote detoxification. 100 mg of glutathione is taken after meals three times a day for a total daily dose of 300 mg. When prescribing glutathione by intravenous drip, 200 mg is usually used. Natural glutathione extracted from a plant and yeast is used for patients who experience a sensitive reaction to synthesized glutathione. Acetylcysteine is used for a similar purpose.

Taurine: Taurine is used for amino acid conjugation. It is taken three times a day for a total daily dose of 3 g.

6.2.2.2 Vitamins

Vitamin C: Vitamin C has many actions. Since it is used also for steroid synthesis, it can help eliminate stress. It has a leading role in detoxification. Vitamin C is a scavenger of superoxide [1]. Superoxide is produced during the decomposition of xenobiotics [2]. Moreover, in the stress state, extensive ingestion of vitamin C is required [3].

One gram of vitamin C is taken after every meal, for a total daily dose of 3 g. If vitamin C is taken before eating, the blood concentration will rise steeply, and the excessive vitamin will be discarded in large quantities in

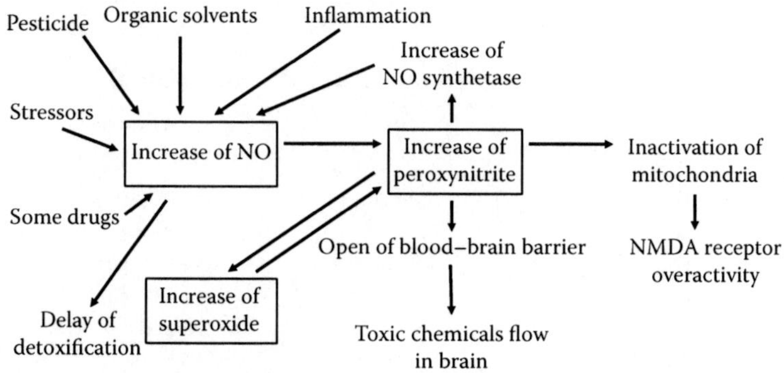

FIGURE 6.2 Schematic illustration of mechanism of CS.

urine without utilization. As vitamin C is acidic, it should be taken after a meal also for protection of stomach walls. There is an upper limit of absorption of vitamin C from the intestinal tract, and the aforementioned quantity is appropriate. In a number of cases, an intravenous drip is used. When prescribing vitamin C via intravenous drip, 10 to 15 g are used.

Vitamin C extracted from rose hips or sago palm rather than synthesized vitamin C is recommended for sensitive patients.

Vitamin E: Vitamin E is lipid soluble and easily passes through the cell membrane. Vitamin C is water soluble and it is difficult for it to enter a cell. Vitamin C reduces the oxidized vitamin E that exits a cell. Simultaneous medication with vitamins C and E is effective in removal of oxidative stress [5].

Vitamin B_{12}, folic acid, multivitamins: Vitamin B_{12} and folic acid have been used experimentally for medical treatment of CS. Pall developed a process of generating peroxynitrite and superoxide following exposure to chemical substances and presented a rationale for the validity of using these medicines [6]. Figure 6.2 shows a simplified version of the Pall mechanism.

Ingestion of coenzyme type vitamin B_{12} is recommended for removal of peroxynitrite. We prescribe 500 μg of methyl cobalamin after every meal. Folic acid is similarly used. The goal of medical treatment of CS is improvement of the condition. Naturally a multivitamin is also recommended. Coenzyme Q_{10} and alpha-lipoic acid are used for improvement of cell function.

6.2.2.3 Symptomatic Therapy

Hypnotics and tranquilizers, if needed, should be prescribed carefully.

6.2.2.4 Minerals

Hair analysis often detects deleterious minerals such as mercury, lead, cadmium, and arsenic in high concentration. The good minerals are in a tug-of-war with the harmful ones.

TABLE 6.1
RDA and ODA of Vitamins and Minerals (Adult)

	RDA	ODA
B_1	1.2–1.4 mg	25–300 mg
B_6	1.2–1.4 mg	25–300 mg
B_{12}	2 μg	25–300 μg
Folic acid	150–200 μg	400–1200 μg
C	60 mg	500–5000 mg
E	8–10 IU	200–800 IU
Ca	1200 mg	1500 mg
Mg	280–350 mg	500–700 mg
Zn	12–18 mg	25–50 mg
Se	13–18 μg	50–400 μg

Magnesium needs to be added for prevention of overactivity of the *N*-methyl-D-aspartate receptor. In addition, magnesium has many pharmacological actions such as relaxation of muscles. Magnesium is effective in helping muscle stiffness or pain and improvement of blood flow.

Zinc is required for removal of superoxides. However, zinc can affect the pancreas. When zinc is administered, measurement of serum amylase must be carried out periodically.

Selenium is required for glutathione metabolism. However, the safety range is very narrow and use of selenium requires caution to avoid overmedication.

The Recommended Daily Allowance and Optimal Daily Allowance of the major vitamins and minerals are shown in Table 6.1.

These values should be adjusted according to weight, health status, and physical constitution.

6.2.2.5 EDTA, Penicillamine, PAM, and Others

Depending on the chemical contaminant, use of a specific antidote is effective. In heavy metal contamination, a chelating agent such as EDTA (ethylenediaminetetraacetic acid) or D-penicillamine should also be taken into consideration.

Organophosphorus compounds are important as causative agents in the onset of CS. Intravenous injection or drip medication of the specific antidote of pralidoxime methiodide (PAM) 1000 mg is an effective treatment. Atropine is used for resolution of a cholinergic state.

6.3 ALTERNATIVE THERAPY

6.3.1 Prayer and Meditation in Zen Buddhism

Fundamentally, prayer, meditation in Zen Buddhism, and yoga promote abdominal breathing. Regular abdominal breathing will produce alpha waves in the brain, as evidenced on EEGs [6], and mental tranquility will result. Mental tranquility is very important and is connected also with some beautiful Christian traditions. Moreover, a song also causes abdominal breathing. Since a song develops from a prayer, this is a natural insight. The back is straightened and a robust song is uttered. Playing a musical instrument also promotes abdominal breathing.

6.3.2 Eastern Medicine and Alternative Medicine

Alternative medicine, such as Chinese medicine, acupuncture, and finger-pressure treatment should also be tried when suitable. Western medicine is very effective for a quick relief of symptoms. However, when it comes to treatment of the cause of chronic disease, it is not always effective. Alternative medicine may be used to treat a patient's subjective symptoms. If the "yin and yang" are out of balance, Chinese medicine may be curative. "Chi kung" was effective in some patients in our clinic. Autogenic training and a self-strengthening method may also be effective.

It seems, however, that there are also individual differences in the responses to alternative medicine.

As noted in Chapter 4, the hippocampus is considered to be a focal point for CS.

The hippocampus is the center of memory and emotion. Changes in blood flow at the hippocampus have been proven by single-photon emission computed tomography (SPECT) at the time of chemical substance load [7]. The nerve cell of the hippocampus is a unique cell in the brain. A nerve cell is renewed at the hippocampus [7]. Moreover, the network of the nerves of the hippocampus is promoted by estrogen [4]. Moreover, the hippocampus is rich in corticoid receptors. The neuron of a hippocampus will be exhausted if the corticoid receptors are continuously stressed. When the memory of a chemical is imprinted through the nose, the memory will always be present. It is necessary to pour pleasant new memories into the brain in as large a quantity as possible and to drive out old unpleasant memories. Shutting oneself up in a room and continuously reviewing old memories stored in the hippocampus should be avoided. Even if something is considered not entirely healthy, it is better for a person to enjoy it and go out to a favorite place.

6.4 NUTRITIONAL INSTRUCTION

The basis of restoration to health is nutrition from a meal. A purified substance called a supplement sometimes shows the same action as a chemical substance. In the allergy experiment described previously, although unrefined sugar did not worsen the allergy, refined sugar did. Although monosodium glutamate makes food more delicious, purified monosodium glutamate is associated with Chinese restaurant syndrome.

Numerous publications are available to provide information on food sources of vitamins and minerals. Vitamins and minerals are abundant in organically grown vegetables.

Here I briefly describe phytochemicals. "Phyto" means a plant in Greek. A phytochemical is a chemical component present in very small quantities that is effective in maintenance of alimentary health. Many different substances are contained in various phytochemicals. Sulfur-containing compounds activate detoxification enzymes. Many phytochemicals have an antioxidant action.

Omega-3 fatty acid, zinc, selenium, magnesium, vitamin D, and B-group vitamins, including B_{12}, may be taken in from food.

ENDNOTES

1. Ishiwata H. 2009. Daily intake of food additives in Japanese and its estimation methods. *Journal of Cookery Science of Japan* 42:198–203.
2. Li G, Hanai Y, Miyata M et al. 1994. Aggravating effects of chloroform and p-dichlorobenzene on experimental allergic conjunctivitis. *Folio Ophtholmol Jpn* 45:475–480.
3. Rapp DJ. 1997. Is this your child's world?: How you can fix the schools and homes that are making your children sick. Bantan Books.
4. Oono K, Miyata M. 1999. Multiple chemical sensitivity. *Allergology and Immunology* 6:990–998.
5. Ishikawa S. 1998. Chemical sensitivity; diagnostic criteria. *Japan Medical J* 3857:25–29.
6. Arambula P, Peper E, Kawakmi M et al. 2001. The physiological correlates of Kundalini Yoga meditation: A study of a yoga master. *Appl Psychophys Biof* 26:147–153.
7. Frank Land PW, Josselyn SA. 2016. Hippocampol neurogenesis and memory clearance [Review]. *Neuropyschopharmacology* 41:382–383.

7 Chemical Sensitivity in Children

Kazuhiko Kakuta, PhD, MD, Director
Kakuta Pediatric Clinic

7.1 GENERAL SYMPTOMS

I here describe an examination of 23 children with sick-building or sick-house syndrome, 15 boys and 8 girls, with an average age of 8.9 years, living in 14 households. They were observed before and after a move to a healthy home. Indoor air chemicals [1] were measured from 1 month to 7 years and 2 months after the move. The symptoms detected were allergic reactions that included 20 cases of bronchial asthmas, 17 of atopic dermatitis, 20 of nasal allergy, 7 of allergic conjunctivitis, 2 of hives, 1 of anaphylaxis, and diverse neurological symptoms. A follow-up evaluation of indoor air quality was made.

The duration of symptoms was divided into three groups:

1. Continuous aggravation: The aggravation of symptoms continues or a new illness appears within 3 years after the move and lasts for 3 months.
2. Temporary aggravation: The aggravation of symptoms continues or any new illness appears within 1 year after the move and lasts less than 3 months.
3. No change: Cases that are not included in the preceding two groups.

Indoor total volatile organic compounds (TVOCs) and aldehydes were measured [2,3]. To determine the influence of organophosphorus (OP) pesticide on the nervous system, the acetylcholinesterase activity of red cells was measured.

7.1.1 Formaldehyde Concentration and Symptoms

In 12 out of 23 cases, "continuous aggravation" was observed. "Temporary aggravation" was observed in 6 cases. In the "continuous aggravation" cases, formaldehyde concentrations were at maximum 0.31 ppm and at least 0.04 ppm. The mean value was 0.21 ppm. In 6 cases of "temporary aggravation," the concentration was at maximum 0.31 ppm, at least 0.06 ppm, and the mean was 0.15 ppm. In 5 cases of "no change," the value was at maximum 0.20 ppm, at least 0.04 ppm, and the mean was 0.12 ppm. Respiratory symptoms were observed in the younger children at the time of the move to a sick house.

7.1.2 TVOC Concentration and Symptoms

Five children with a high TVOC concentration showed aggravation of respiratory tract symptoms. The TVOC concentrations of four out of five cases were higher than 3000 μg/m^3 and are classified in the continuous aggravation group of respiratory tract symptoms. One of them showed sick-building syndrome.

7.1.3 Red Cell Cholinesterase

In 8 cases out of 17 cases, the red cell cholinesterase (RChE) was 1.5 units or less, meaning a reduction. Exposure to an OP pesticide is considered to be a cause of this reduction. It is known that acetylcholine receptor exists in a lymphocytic leukocyte (thymus-derived T lymphocyte) [4]. OP compounds probably have a large impact on the immune system as well as the nervous system.

In recent years, the number of children with allergies has been continuously increasing. It is believed that environmental chemicals influence the immune, nervous, and endocrine systems in addition to the respiratory system [1,2,5].

The leading symptoms in a younger group were seen in the mucous membrane of the respiratory tract and the skin. Nervous system symptoms consisting mainly of headache, nausea, vomiting, lightheadedness, and vertigo increased with age [5]. Sakamoto and others have reported an increase of airway secretions that strongly coincided with formaldehyde inhalation [1,6], which induces a cough. The neuropeptide substance P is secreted from the C-fiber terminal. It stimulates mast cells and an allergy may be induced. Sakamoto and others have reported that toluene produces this state also [7]. Moreover, in an experiment in which murine mast cells and murine superior cervical ganglia cells were co-cultivated, the transmission of a neural excitation and an excitation of a mast cells was bidirectional. It is reported that the transmission of excitation to a mast cell from a neuron is mediated by substance P [8]. A stimulation of the sensory neuron of the mucous membrane induces an allergy and an allergic reaction induces neural excitation, thus possibly creating a vicious cycle [9,10]. The continuous worsening of a symptom such as a cough or asthma that occurred after children moved to new housing may be based on such a vicious cycle in the mucous membrane of the respiratory tract. There is a close similarity between these children's symptoms and those of a syndrome like the asthma that Millqvist and others have reported in adults [11].

Much attention should be paid to infants. In a study by Ruda et al., after the soles of newborn rats were subjected to a strong inflammatory stimulus, hypersensitivity to pain remained even after growth [12]. This experimental result indicates a probability of inducing CS from hypersensitivity of the nervous system after growth, following exposure to a strong environmental chemical stimulus in infancy.

It has been observed that the aggravation of allergy and acceleration of antibody production are induced by formaldehyde [13]. As the effects of trace chemicals in the environment are much entangled, we should be careful in daily life.

7.2 EFFECTS OF INDOOR CHEMICALS ON THE INTELLIGENCE AND COGNITIVE FUNCTION OF CHILDREN

I have reported that indoor chemicals may change cerebral blood flow, and that those changes may be related to a neuropeptide or serotonin [14–17]. Indoor chemicals may change neural excitability and may influence the brain development of infants.

Eleven children with sick-house syndrome and their siblings were examined. They moved to a new home from the sick house after diagnosis of sick-house syndrome. Three of them were born after the move and the other eight ranged in age from 2 years 11 months old to 12 years old at the time of move.

There was no problem in adaptation to their school life. Their lifestyle behavior was appropriate for their age. But five of six children in elementary school upper classes had some problems in the writing of Chinese characters, the literacy of Chinese characters, the English spelling, and map reading.

7.2.1 Evaluation of Intelligence

The Wechsler Intelligence Scale for Children-Third Edition (WISC-III) was used for the evaluation of intelligence. WISC-III is an intelligence test for children from 5 years 0 months old to 16 years 11 months old. The test comprises the Verbal Intelligence Quotient (VIQ), Performance Intelligence Quotient (PIQ), and Full-Scale Intelligence Quotient (FIQ). The mean intelligence quotient for each age was set at 100, and the standard deviation was 15.

VIQ was 107.3 ± 6.9 and all subjects were within the normal range. On the other hand, PIQ was 93.5 ± 13.6. A statistically significant difference was observed between the PIQ mean and VIQ mean (t-test, $p < 0.01$) (Figure 7.1).

A learning disturbance is considered when the difference between VIQ and PIQ (VIQ – PIQ) is 15 points or more. In five children, the differences were 18, 20, 26, 31, and 35 respectively, and they were diagnosed as having a learning disturbance. As performance IQ was lower than standard, its subsets were analyzed.

The subtest items with low scores on the PIQ were Coding (mean 8.4 ± standard deviation 2.6), Object assembly (8.7 ± 2.2), Symbol search (8.9 ± 3.3), and Mazes (7.6 ± 4.2) as shown in Figure 7.2. In the other subtest items, scores were within normal limits.

In a pictures completion test, a subject looks at a picture card and answers questions about the area that is missing in the picture, either by pointing or speaking. This time, a card showing an orange was used (Figure 7.3).

A large reduction in the mean was not found. However, the number of correct answers in the question plates "orange" was only 3 among 11 persons.

7.2.2 WISC-III Test Results and Concentration of Indoor Chemicals

A reverse correlation was observed between the correct answer on the "orange" and the indoor p-dichlorobenzene concentration. In the correctly answered cases, the

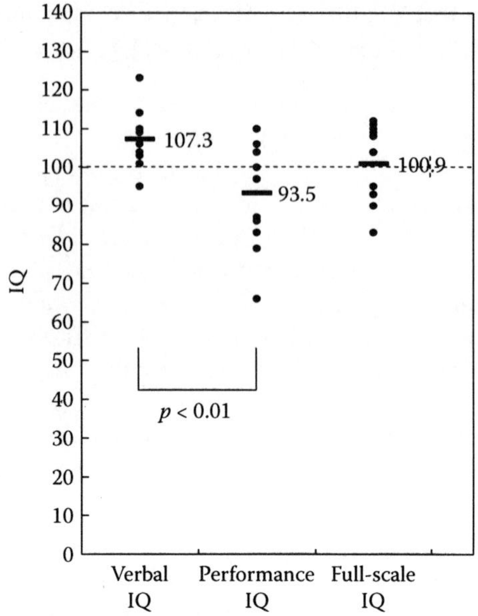

FIGURE 7.1 Intelligence evaluation. VIQ, PIQ, and FIQ.

p-dichlorobenzene concentration was at a low value, and in the incorrectly answered cases it was at a high value (Figure 7.4).

Moreover, a significant relationship was observed between PIQ and indoor *p*-dichlorobenzene concentration as well as formaldehyde concentration ($p < 0.05$).

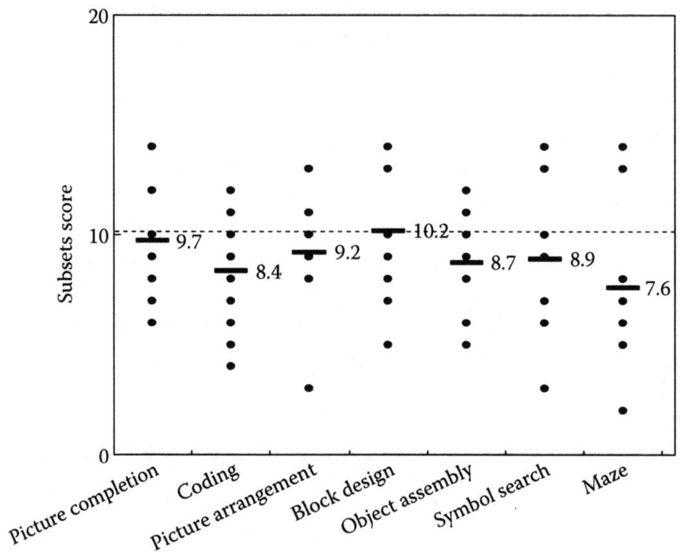

FIGURE 7.2 Subset scores of PIQ.

FIGURE 7.3 The divisions of three segments are missing in an orange. This is a mimicry of a figure in the WISC-III test.

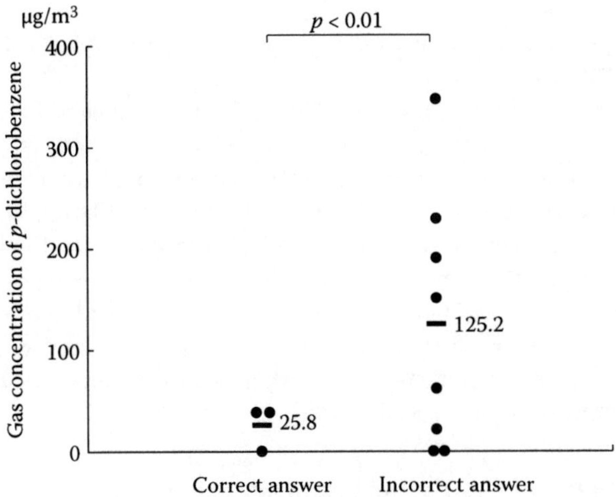

FIGURE 7.4 Correct and incorrect answer for the "orange" figure and gas concentration of *p*-dichlorobenzene.

7.2.3 Development of Sick-House Syndrome

The subject was born in a newly built house where an OP pesticide was used under the floor for termite extermination. Cough and wheezing started appearing 2 months after birth, and improvement was not observed. Measurements of indoor air pollution showed a very high level of TVOCs, 6043.5 μg/mL, and of *p*-dichlorobenzene, 663 μg/mL. RChE fell to 1.4 units (normal value, 1.8 to 2.2 units) and OP exposure was also suspected. His VIQ was a normal 101 at 5 years and 2 months. His PIQ fell to 66, and he was not able to a draw a picture of a person in a manner typical for his age. After medical treatment as well as the environmental improvement of the house the symptoms improved. PIQ was normalized at 97, and at 7 years of age he was able to describe the image of a person whose picture he was shown (Figure 7.5).

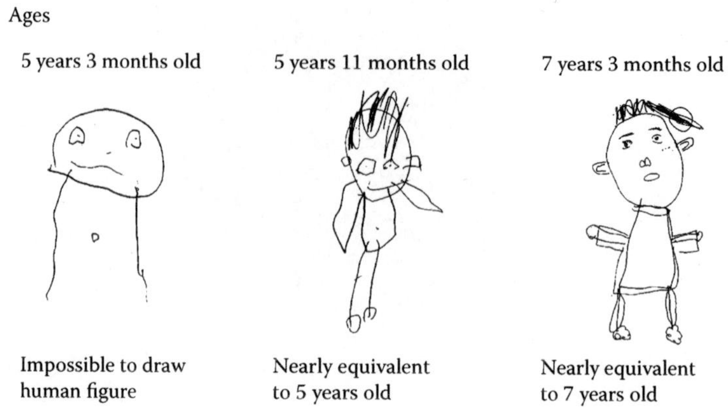

FIGURE 7.5 Drawings of a human figure.

In the present examination, indoor chemicals had strong tendency to influence the PIQ on the WISC-III.

Various kinds of chronic toxicity of organochlorine compounds and OPs are known. An organochlorine pesticide such as dichlorodiphenyltrichloroethane induces a disturbance of the sodium channel that exists in the neural membrane, producing continuous excitement of the nerve [18]. *p*-Dichlorobenzene is also an organochlorine compound and it appears to have the same effect [19,20]. Guillette and others investigated preschool children of the Yaqui Valley zone in Mexico. They compared the pesticide-exposed group living near a farm and another, unexposed, group living a long distance from a farm. The pesticide-exposed group performed poorly on drawing or visual memory [21]. Much attention should be paid to the contaminated chemicals in the air as they affect children.

7.3 CASE REPORT OF CS IN A YOUNG BOY WHO HAD BEEN TREATED WITH SUMATRIPTAN SUCCINATE AND SELECTIVE SEROTONIN REUPTAKE INHIBITORS

A patient in my clinic, a 9-years-old boy suffered from an OP used to exterminate termites. Immediately after moving into the home, neurologic manifestations, headache, nausea, vertigo, and a feeling mimicking intoxication appeared. RChE was low 1.5 (normal value 1.8 to 2.2 units). An infrared pupillometer test, revealed a significantly small pupillary area, indicating parasympathetic nervous superiority caused by OP compounds used to exterminate termites. After ventilation of the home environment and dietary therapy, the symptoms improved to some extent. At the age of 14 the subject experienced symptoms associated with chemical exposure, including throbbing headache, nausea, lightheadedness, and rash. Subcutaneous injection of sumatriptan (Imigran®) (serotonin receptor stimulating agent) reduced his headache and the rash subsided within 30 minutes, as shown in Figure 7.6. It was thought that abnormal expansion of blood vessels might be induced in the brain and the skin. The vasoconstrictive effect of sumatriptan might cause the nervous system and skin symptoms to subside.

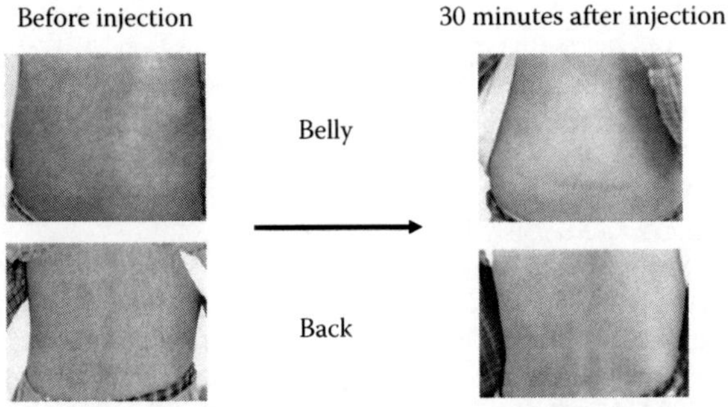

FIGURE 7.6 The rash was reduced 30 minutes after administration of sumatriptan.

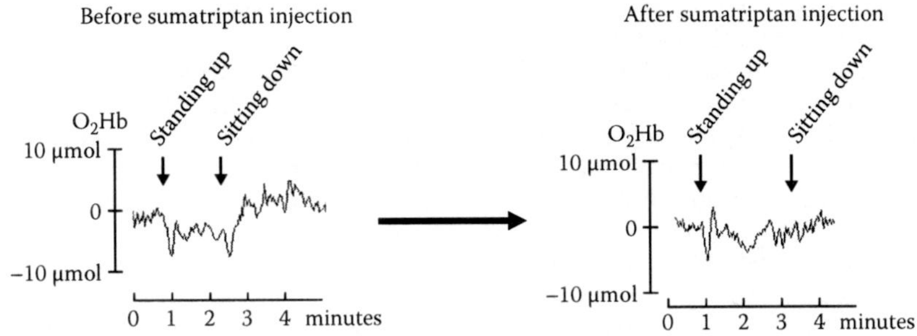

FIGURE 7.7 Oxygenated hemoglobin at frontal area of brain measured by NIRS. Fluctuation of cerebral blood flow can be seen in standing up and sitting down motions, which became stable after injection of sumatriptan.

The same symptoms appeared later. An examination using near-infrared spectroscopy (NIRS) was performed. The cerebral blood flow of patients with CS is usually unstable and fluctuates with gas inhalation or a standing up position (refer to NIRS in Chapter 5) [16,22]. Worsening of the cerebral blood flow that appeared in the standing up trial was improved with sumatriptan injection and thereafter the symptoms subsided (Figure 7.7).

Selective serotonin reuptake inhibitors (SSRIs) inhibit serotonin reuptake in a nerve cell, increasing the serotonin concentration in the extracellular fluid. SSRIs were expected to have the same effect as sumatriptan. An NIRS examination in the standing-up position after administration of an SSRI (fluvoxamine maleate) for about 2 months showed marked improvement.

ENDNOTES

1. Kakuta K et al. 2004. The influence that the indoor chemical substance pollution gives to an onset and development of an allergic disease. *Jpn J Clin Ecol* 13:26–34.
2. Yoshino H et al. 2002. Field study for sick house: Relation between Indoor air qualities and occupants' health conditions. *Neuro-ophthalmol Jpn* 19:188–200.
3. Iida N et al. 2002. Field study about residential environment and health condition in sick house. *Jpn J Clin Ecol* 11:77–87.

4. Toyabe S, Iiai T, Fukada M et al. 1997. Identification of nicotinic acetylcholine receptors on lymphocytes in the periphery as well as thymus in mice. *Immunology* 92:201–205.
5. Ito K et al. 1996. Role of tachykinin and bradykinin receptors and mast cells in gaseous formaldehyde-induced airway microvascular leakage in rats. *Eur J Pharmacol* 307:291–298.
6. Sakamoto T, Miyake M. 2010. Health effects of formaldehyde, as an indoor air pollutant. *Kaibogaku Zasshi* 85:35–41.
7. Sakamoto T, Miyake M, Ito A et al. 2000. The effect of single dose inhalation of toluene on the permeability of vessels of air way. *Jpn J Allergol* 49:984.
8. Suzuki R, Furuno T, Mckay DM et al. 1999. Direct neurite-mast cell communication in vitro occurs via the neuropeptide substance P. *J Immunol* 163:2410–2415.
9. Meggs WJ. 1999. Mechanisms of allergy and chemical sensitivity. *Toxicol Indust Health* 15:331–338.
10. Bascom R, Meggs WJ, Frampton M et al. 1997. Neurogenic inflammation: With additional discussion of central and perceptual integration of nonneurogenic inflammation. *Environ Health Perspect* 105:531–537.
11. Millqvist E, Bende M, Lowhagen O. 1998. Sensory hyperreactivity—A possible mechanism underlying cough and asthma-like symptoms. *Allergy* 53:1208–1212.
12. Ruda MA, Ling QD, Hohmann AG et al. 2000. Altered nociceptive neuronal circuits after neonatal peripheral inflammation. *Science* 289:628–630.
13. Tarkowski M, Gorski P. 1995. Increased lgE anti-ovalbumin level in mice exposed to formaldehyde. *Int Arch Allergy Immunol* 106:422–424.
14. Kakuta K. 2003. Sick house syndrome and sick school syndrome: From the view point of pediatrician. *Allergy Immunol* 10:1595–1604.
15. Kakuta K, Yoshino H, Amano K et al. 2004. Sick house syndrome in child hood. *Jpn J Clin Ecol* 13:85–92.
16. Kakuta K, Yoshino H, Amano K et al. 2003. A chemical gas short time inhalation examination and an orthostatic stress test before and after gas inhalation using near infrared spectroscopy in order to diagnosis of sick house syndrome. *Jpn J Clin Ecol* 12:15–26.
17. Kakuta K, Hojo S, Yoshino H et al. 2002. Influence of chemical materials on adolescent children with allergic diseases. *Neuro-ophthalmo Jpn* 19:176–187.
18. Vijverberg HPM, van der Zalm JM, van den Bercken J. 1982. Similar mode of action of pyrethroids and DDT on sodium channel gating in myelinated nerves. *Nature* 295:601–603.
19. Coats JR. 1990. Mechanisms of toxic action and structure-activity relationships for organochlorine and synthetic pyrethroid insecticides. *Environ Health Perspect* 87:255–262.
20. Narahashi T. 1996. Neuronal ion channels as the target sites of insecticides. *Pharmacol Toxicol* 79:1–14.
21. Guillette EA, Meza MM, Aquilar MG, Garcia IE. 1998. An anthropological approach to the evaluation of preschool children exposed to pesticides in Mexico. *Environm Health Perspect* 106(6):347–353.
22. Kamiyama M, Kakuta K, Yoshino H, Hojo S, Ishikawa S. 2003–2005 fiscal year summed report. Proceedings of 2003–2005, Committee on Sick House Syndrome, Chairman Ishikawa S, supported by Ministry of Health, Labor and Welfare, Japan.

8 Multiple Chemical Sensitivity—Medical Aspects from Germany

Klaus-Dietrich Runow, MD, Director
Diagnostic and Therapy Center, Institute for Functional and Environmental Medicine (IFU), Germany

Since September 8, 2009, 50 million chemical substances have been registered with the Chemical Abstracts Service (CAS), which is a part of the American Chemical Society. The CAS registers 3.000 new substances daily. Although these represent only registered chemical substances, in modern life a person is confronted with 80.000 chemicals in his or her daily life. A growing number of people are not able to tolerate this amount of chemicals and develop allergies and multiple chemical sensitivity (MCS), which can occur suddenly (high dose initiated) or after exposure to small amounts of chemicals, drugs, odors, and so forth over a long time (low dose initiated).

The end "product" of chemical exposure is cancer—a metaphor of our overstimulated civilization. Cancer is the *number one environmental disease.* A Scandinavian study analyzed 44,786 pairs of twins in Sweden, Denmark, and Finland. The researchers examined cancer at 20 sites including stomach, colon, lung, breast, and prostate. The conclusion: "The finding indicates that environment has the principal role in causing sporadic cancer" (almost 70%) [1].

If two-thirds of cancers are caused by environmental factors, then we have the choice to change the environmental attacks on our genes. We have to think about the food we eat, which daily chemicals we use, which materials we choose for building a house, and so forth. Of course, the majority of people on earth don't have a choice: they have to work in toxic offices and industrial facilities. Therefore it is a primary task for persons in leading positions such as managers, physicians, politicians, and so forth to make preventive decisions. The current production methods and ways of living carry too high a cost, as they will destroy the brains and immune systems of whole generations.

In China so-called "cancer villages" exist near industrial facilities. This is a reason for great concern, because this message came from the official government! We will, however, observe a growing number of cancer cases in the future in other areas, not only in the extremely polluted country of China.

It is expected that in Europe every second (!) man and 43% of women will receive a cancer diagnosis during their life time [2,3].

8.1 CHEMICAL SENSITIVITY

> Our science looks at a substance-by-substance exposure and doesn't take into account the multitude of exposures we experience in daily life. If we did, it might change our risk paradigm. The potential risks associated with extremely low-level exposure may be underestimated or missed entirely.
>
> **Heather Logan**
> *Canadian Cancer Society*

People who suffer from CS react to very small amounts of many commonly used toxic chemicals. Many chemicals have the potential to act as toxicants in initiating CS [4]:

Pollutants in buildings (new: nanoparticles)
Pollutants in cars (fogging substances)
Emissions from plastic materials
Solvent agents
Perfume
Cosmetics
Laundry powder
Smoke (from cigarettes and chimneys)
Inhalants (pollen)
Foods
Drugs

Mainstream physicians often argue, "Patients who suffer from CS are only afraid of chemicals and pollutants." They explain to CS patients: "CS may exist. But you don't have it. Your symptoms are psychosomatic."

Functional medicine, that is, applied environmental and nutritional medicine, moves away from a *disease-centered* to a *patient-centered* approach [5]. Sir William Osler stated, "It is much more important to know what sort of a patient has a disease than what sort of disease has a patient" [5]. Try to understand what your patient's health and life were like before the onset of the present illness.

8.2 BIOCHEMICAL INDIVIDUALITY

CS patients have an individual biochemical makeup with individual triggers and/or mediators that have led to chronic inflammation causing fire in body and brain. Once activated, chronic inflammation can lead to an immune activation of our glia cells in the brain—the so-called blood–brain barrier. Because glia cells are associated with the immune cells in the gut, we can understand the sequelae of chronic digestive disorders in the brain: mood and behavioral disturbances, epilepsy, migraine, hyperactivity, depression, and so forth. The causes of inflammations in the gut and digestive system can be the intake of drugs, chemicals, solvents, heavy metals, infections, and hidden food allergies.

A recent animal study demonstrated that mediators released from mast cells in tissue taken from patients with irritable bowel syndrome resulted in increased firing of visceral-nociceptive sensory neurons (pain-triggering neurons) in rats [6,8]. What this means is that inflammatory mediators influence the nervous system. In addition, there is a link

TABLE 8.1
Gastrointestinal Function Profile

Pathogens			
Bacterial Pathogens	**Result**	**Genotype**	**Expected**
Campylobacter	Negative		Neg
C. difficile Toxin A	Negative		Neg
C. difficile Toxin B	Negative		Neg
E. coli O157	Negative		Neg
Enterotoxigenic *E. coli* LT	Negative		Neg
Enterotoxigenic *E. coli* ST	Negative		Neg
Shiga-like Toxin *E. coli* stx1	Negative		Neg
Shiga-like Toxin *E. coli* stx2	Negative		Neg
Salmonella	Positive		Neg
Shigella	Negative		Neg
Vibrio cholera	Negative		Neg
Yersinia enterocolitica	Negative		Neg
Parasitic Pathogens			
Cryptosporidium	Negative		Neg
Entamoeba histolytica	Negative		Neg
Giardia	Negative		Neg
Viral Pathogens			
Adenovirus 40	Positive		Neg
Adenovirus 41	Positive		Neg
Norovirus GI	Negative		Neg
Norovirus GII	Negative		Neg
Rotavirus A	Negative		Neg
H. pylori			
Helicobacter pylori	4.03 E7	High	<7.0 E3
Virulence Factor, cagA	Positive		Neg
Virulence Factor, vacA	Negative		Neg
Normal Bacterial Flora			
Bacteroides fragilis grp	6.2 E7		5.0 E5–3.2 E9
Bifidobacter	9.5 E7	Low	>8.9 E9
Enterococcus	2.0 E7	High	1.2 E4–3.1 E6
E. coli	4.2 E8	High	1.0 E4–7.6 E7
Lactobacillus	4.4 E5		1.0 E6–5.8 E9
Opportunistic Bacteria			
Potential Autoimune Triggers			
Citrobacter spp.	8.1 E6	High	<1.0 E4
Klebsiella pneumoniae	4.2 E3		<7.2 E3
Proteus spp.	1.2 E3		<6.2 E3
Proteus mirabilus	1.7 E3		<1.0 E3
Yersinia enterocolytica (from pg 1)	Negative		Neg

(*Continued*)

TABLE 8.1 (CONTINUED)
Gastrointestinal Function Profile

Bacterial Pathogens	Result	Genotype	Expected
Additional Dysbiotic/Overgrowth Bacteria			
Morganella morganii	8.0 E2		<1.0 E3
Pseudomonas spp.	1.2 E3		<2.5 E3
Pseudomonas aeruginosa	5.3 E8	High	<1.0 E3
Staphylococcus spp.	9.2 E3		<1.0 E4
Streptococcus spp.	4.2 E2		<1.0 E3
Parasites			
Blastocystis hominis	9.1 E5	High	0.00
Dientamoeba fragilis	Negative		Neg
Endolimax nana	Negative		Neg
Entamoeba coli	Negative		Neg
Chilomastix mesnelli	Negative		Neg
Pentatrichomonas hominis	Negative		Neg
Microsporidia spp.	Negative		Neg
Fungi/Yeast			
Candida albicans	8.7 E7	High	>5.0 E3
Candida spp.	Negative		Neg
Cyclospora cayetanenensis	Negative		Neg
Geotricum spp.	<1000 cfu/g	Low	Neg
Trichosporon spp.	Negative		Neg
Additional Tests			
SIgA	458	High	510–2040 ug/mL
Anti-gliadin	0.8		0.0–1.1 ug/g
Elastase 1	102	Low	>175 mcg/g
Lactoferrin	3.8		0.0–7.2 ug/mL
Occult blood	Negative		neg
Antibiotic Resistance Genes			
	Phenotype	**Genotype**	**Expected**
Salmonella			
Sulfonomides	Negative	Positive	Neg
Trimethoprim	Negative	Negative	Neg
Fluoroquinolones	Negative	Negative	Neg
Macrolides	Negative	Negative	Neg
Pseudomonas			
Sulfonomides	Negative	Positive	Neg
Trimethoprim	Negative	Negative	Neg

between the hypothalamic–pituitary–adrenal (HPA) axis and the gut immune system. Chronic stress with chronic high levels of adrenocotropic hormone and cortisol cause the release of the proinflammatory cytokines interleukin-6 (IL-6) and interleukin-8 (IL-8).

Because chemicals are able to upregulate inflammatory processes in the body and also in the brain, both origins of inflammation have to be analyzed in CS patients.

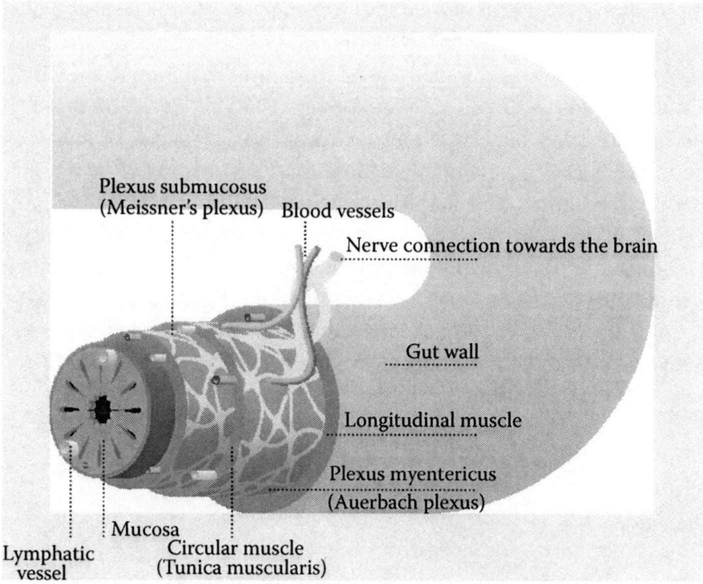

FIGURE 8.1 The intestinal wall contains 100 million nerve cells. Ninety percent of the nerve cells go up to the brain; only 10% of nerve cells lead down from the brain.

Given the central immunological and neurological role of our gut, we have decided to recommend the latest stool test technology. DNA analysis is the new gold standard for identification of bacteria in clinical microbiology and has greatly facilitated the identification and classification of intestinal microbiota composition [7–17].

One of my patients suffering from CS showed the stool test results presented in Table 8.1. Intestinal dysbiosis, namely overgrowth of yeast (Candida), Parasites (Blastocystis), Helicobacter pylori and bacteria associated with inflammatory and autoimmune diseases,was recognized.

More than 100 million nerves from the enteric plexus lead directly into the brain. Ninety percent of the nerve fibers go into the brain—only 10% lead from the brain to the gut (Figure 8.1). Based on this knowledge, we have to investigate whether CS patients have developed chronic inflammatory processes in their body—a continuous fire caused by toxic or biological triggers and mediators.

Chemicals such as ingested heavy metals disturb the digestive system by, for example, blocking enzymes such as peptidase, which is responsible for digesting proteins in our food. If these enzymes are lacking, the gut bacteria produce putrefactive short-chain fatty acids that can cause severe symptoms in the gut and also in other tissues.

8.3 RANDOLPH'S SPECIFIC ADAPTATION SYNDROME IN RESPONSE TO ENVIRONMENTAL STRESSORS

The term "stress" was first used by psychologists before the endocrinologist Hans Selye adopted it in the 1930s. He later broadened and popularized the concept to include the response of the body to any demand. In Selye's terminology, "stress" refers to a condition and "stressor" to the internal reaction causing stress.

Stress covers a huge range of phenomena from mild irritation to the kind of severe problems that might result in a real breakdown of health. Signs of stress (including from chemical stressors) may be cognitive, emotional, physical, or behavioral. They include poor judgment; a generally negative outlook; excessive worrying, moodiness, irritability, agitation, and inability to relax; feeling lonely or isolated; depression; aches and pains; diarrhea or constipation; nausea; dizziness; chest pain; rapid heartbeat; eating too much or not enough; sleeping too much or not enough; withdrawing from others; procrastinating or neglecting responsibilities; using alcohol, cigarettes, or drugs to relax; and nervous habits (e.g., nail biting or pacing).

Selye exposed rats and other animals to unpleasant or harmful stimuli and discovered three stages of adaptation:

1. Alarm
2. Resistance
3. Exhaustion

There are always the same stress response patterns (activation of the HPA axis, production of cortisol, and so on) to different stressors, which Selye called the general adaptation syndrome (GAS). After a phase of apparent adaptation (phase 2), for example, to chemicals, food, physical stress, and so on the person reaches a state of exhaustion: a breakdown of biochemical systems in the body. The rats in Selye's trial (exposed to the stressor coldness) died from adrenal hypofunction—the adrenals had been completely destroyed in phase 3.

In comparison to Selye's GAS, the environmental physician Prof. Theron Randolph (an internist and psychiatrist) described the *specific adaptation syndrome*

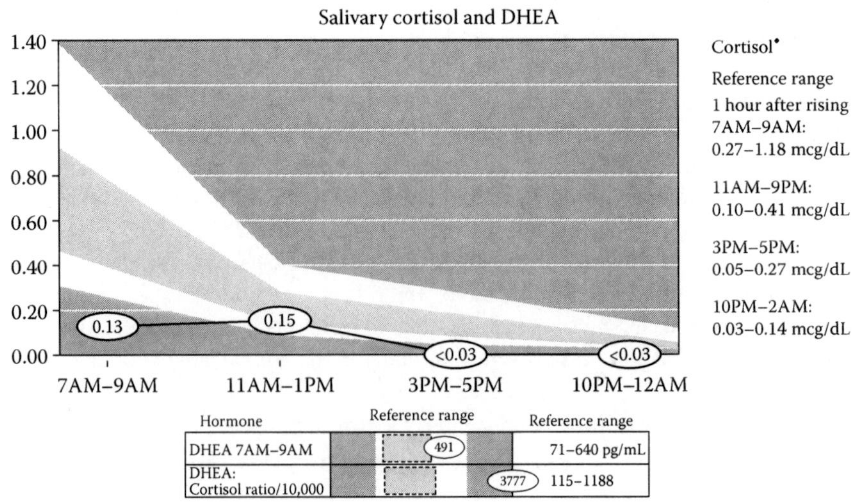

FIGURE 8.2 Adrenal hypoactivity (cortisol "burnout").

(SAS). He discovered that his CS patients had an individually long period of an *apparently* (not really!) adaptation phase (resistance phase 2) without symptoms in spite of chemical exposure. At the end, in phase 3 (exhaustion) the person develops severe acute and chronic reactions such as CS.

The phase of exhaustion can be described as "burnout syndrome": Patients show physical and neurological symptoms even after contact with low doses of chemicals, drugs, perfume, and also natural substances (e.g., terpenes). As shown in Figure 8.2, hormonal exhaustion is detected—a dangerous situation. The lack of cortisol makes this person extremely vulnerable. An ordinary bout of flu can lead to death because the person is not able to generate stress protection through cortisol.

8.4 CAUSES: FIRE IN THE BODY AND THE BRAIN

We here discuss three factors implicated in causation of CS: precipitating events, triggers, and mediators.

1. *Precipitating events*: A reactive arthritis after an amebic infection (parasites) is called a precipitating event. In CS patients, chemical overexposure (to solvents/paintings, pesticides, antibiotics, heavy metals etc.) can be the precipitating event.
2. *Triggers*: Parasites, pathogenic bacteria, yeast and molds, food antigens, and ingested toxins can act as continuous triggers of inflammations such as chronic colitis and digestive disorders. They have to be diagnosed and eliminated. In CS patients, the triggers are different chemicals such as solvents, plastic materials, perfume, pharmaceuticals, and also foods and food additives. They have to be diagnosed and eliminated.
3. *Mediators*: Mediators are intermediates/metabolites that contribute to the manifestations of disease. Like triggers, mediators do not in and of themselves cause disease. Mediators vary in form and substance; they may be biochemical (cytokines, prostanoids), ionic (e.g., hydrogen ions), social (e.g., positive reinforcement for staying ill), psychological (e.g., fear), or cultural (e.g., beliefs about the nature of illness). Common mediators of illness include hormones (adrenal hypofunction, imbalances of melatonin, progesterone etc.), free radicals, fear of pain or loss, poor self-esteem, learned helplessness, and lack of relevant health information.

The striking characteristic of all mediators is their *lack of disease specificity*.

8.5 DIAGNOSTIC PROGRAM RECOMMENDED FOR CS PATIENTS

8.5.1 Metabolic and Nutritional Status (Organic Acid Profiles), Toxins

The most productive area in biochemical research over the past few years has been the identification and isolation of biochemical disease mediators. In addition to the usual medical analysis including toxicological tests, the new urine test—organic acid analysis (Organix Profile)—can help discover triggers and mediators in CS patients. The original Comprehensive Organix Test (patent pending) provides a view into the body's cellular metabolic processes and the efficiency of metabolic function. Identifying metabolic blocks or problems with detoxification, gut dysbiosis, or oxidative stress that can be treated nutritionally allows us to tailor interventions. Targeted treatment can help maximize patient responses and lead to improved patient outcomes.

Patients with CS often show mitochondrial disturbances that cause a lack of energy (ATP) production. Because the body needs about 80% of the energy for detoxification purposes, mitochondrial dysfunction leads directly to an insufficient elimination of toxins. Substances such as glutathione, alpha-lipoic acid, and coenzyme Q_{10} (ubiquinone) can help support the mitochondria, increasing the energy level and the detoxification capacity.

8.5.2 Toxic Metals

Some elements can accumulate in tissues, causing toxic effects. Metal toxicity is a significant environmental health concern. A toxic load of aluminum, lead, cadmium, mercury, or arsenic is capable of rendering considerable damage to the brain and nervous system, particularly in children. Toxic elements produce their many negative effects through various mechanisms.

One mechanism, irreversible enzyme inhibition, is illustrated by the anemia caused when lead binds to enzymes in the hemoglobin synthesis pathway. Mercury causes enzyme poisoning. Lead and mercury reduce glutathione and cause a vitamin B_{12} deficiency (methyl-B_{12}).

The cancer-inducing effect of arsenic seems to be due to an inhibition of DNA repair. Genotoxicity, in which chromosomes are damaged, is linked to the free radical generation capacities of cadmium, lead, and nickel. Nutritional and toxic elements can be measured in blood, urine, and hair.

Hair has a long history of successful use in detecting chronic exposure to toxic metals in human and animal models because hair concentrates heavy metals several hundred fold above concentrations found in blood. When any of the toxic heavy metals are elevated in hair, there is reason to investigate the origin of exposure. High levels in hair may reflect early chronic exposure before other signs and symptoms appear. Physicians are becoming more and more concerned about the toxicity of aluminum, which is used in multiple products in our daily lives, for example, deodorant (antiperspirants), vaccinations, antacids, and products used for filtering drinking water. Aluminum is considered a carcinogenic and neurotoxic element (increasing cases of Alzheimer's disease and other neurodegenerative disorders). In my opinion, we have to eliminate aluminum immediately from our daily lives. The University of Vienna uses aluminum to induce all kinds of allergies in animals for research purposes. This fact shows that aluminum as a trigger of a broad spectrum of allergies is one reason for the rising number of allergies and also neurological symptoms such as attention deficit hyperactivity disorder (ADHD) in our society. In Germany the insurance company Allgemeine Ortskrankenkasse (AOK) published in February 2013 the finding that the number of ADHD children doubled in only two years!

Another diagnostic tool is the analysis of *porphyrins*. The porphyrins profile can help identify the severity of heavy metal toxicity or organic chemical exposure in patients. Chemical exposure and a heavy toxic burden can have physiological effects resulting in impaired metabolism and cellular function. Porphyrin testing in the urine also helps to monitor the treatment efficiency in patients suffering from MCS and neurological symptoms.

8.5.3 Phthalates and Parabens Profile

The phthalates and parabens profile can help identify everyday exposures to toxins from the use of items such as personal care products and plastic food containers. Environmental toxins should be evaluated as a "first step" to help patients get back on the road to wellness.

8.5.3.1 Why Assess Phthalate and Paraben Levels?

Exposure to phthalates and parabens is more common than one may realize. Phthalates and parabens are often classified as xenoestrogens, foreign compounds in the body functioning as endocrine disruptors by binding specifically to estrogen receptors.

Endocrine disruptors are associated with conditions such as

- Endometriosis
- Infertility
- Breast cancer
- Ovarian cancer
- Prostate cancer
- Testicular cancer
- Decreased sperm count

Other health problems associated with daily exposures are

- Liver toxicity
- Immune effects such as allergies and asthma
- Reproductive toxicity
- Pubertal development

8.5.3.2 Where Are Phthalates and Parabens Found?

Phthalates, also called "plasticizers," are found in numerous everyday products such as

- Children's toys
- Cosmetics
- Cleaning products
- Air fresheners
- Perfumes
- Furniture
- Vinyl flooring
- Plastic food containers
- Medical products

Di-(2-ethylhexyl) phthalate ester (DEHP) is a common additive to polyvinyl chloride (PVC). This additive helps make PVC soft and pliable to be molded into eye-pleasing shapes. PVC products are marked with the plastic identification code 3. The analytes measured in this profile are metabolites of DEHP. In perfumes and air fresheners, phthalates are often listed as "fragrance."

8.5.4 Bisphenol A

Bisphenol A (BPA) has been in use for more than 50 years and belongs to the group of hormone-disrupting chemicals. It is a phenolic chemical used in the manufacture of polycarbonate plastics and epoxy resins. While air, dust, and water are possible sources of exposure, BPA in food and beverages containers accounts for the majority of daily human exposure. It is well absorbed orally. BPA leaches into food via packaging and during agricultural production. Storage temperatures are a primary influence on the degree to which BPA leaches from polycarbonate bottles. In a review of the 2005–2006 National Health and Nutrition Examination Survey (NHANES) consumption data study, school lunches and meals outside the home were statistically associated with higher urinary BPA levels. Research also found increased BPA in canned vegetables, while BPA concentrations did not vary according to consumption of fresh fruits and vegetables, canned fruit, or store-bought fresh and frozen fish [18–20]. "The evidence for adverse reproductive outcomes (infertility, cancers, malformations) from exposure to endocrine-disrupting chemicals is strong, and there is mounting evidence for effects on other endocrine systems, including thyroid, neuroendocrine, obesity and metabolism, and insulin and glucose homeostasis" [21].

Table 8.2 shows our new toxic chemical panel from Genova Diagnostics/USA including solvents, BPA, phthalates, and parabens and pesticides.

8.5.5 Hidden Food Allergies

CS patients very often suffer from food allergies. IgG_4 antibodies are associated with nonatopic or "delayed" food reactions that can worsen or contribute to many different health problems. These reactions are considered the most common form of immunologically mediated food intolerance. An IgG_4 response to food is actually more common than the IgE response, which causes an immediate reaction. IgG_4 acts as a blocking antibody, protecting the individual from potentially fatal IgE reactions. These reactions are more difficult to notice because they can occur hours or even days after consumption of an offending food. In some cases, a person's reaction to a food may occur several days after eating the offending food and he or she may not recognize a connection between the food and his or her symptoms. These "hidden" food allergies are caused by increasing blood levels of IgG_4 antibodies in reaction to specific foods. A new patent-pending IgG_4 assay allows our test to show fewer false positives, leading to better patient compliance. In addition to the antibody tests we recommend the lymphocyte transformation test (LTT) on food antigens. Living white blood cells are brought into contact with a group of 25 of the most common food antigens. A high cellular stimulation caused by food shows inflammation—a late immunological response. This means that the patients don't have immediate reactions after

TABLE 8.2
Toxic Effects Core (Chemical Occurrence and Related Exposure)

Category	Alkylphenols	Organochlorines	Organophosphates	Plasticizer/Preservatives	Polychlorinated Biphenyls (PCBs) (Special Organochlorine Group)	Volatile Organic Compounds (VOCs)
Individual Test Profile Name	Bisphenol A	Chlorinated Pesticides	Organophosphates	Phthalates and Parabens	Polychlorinated Biphenyls (PCBs)	Volatile Solvents
All Compounds Tested For In The Complete Toxic Effects CORE Profile	BPA	DDE	DMP	Phthalates	Dioxin-Like	Benzene
	Triclosan	DDT	DMTP	MEtP	PCB 77	Ethylbenzene
	4-Nonylphenol	Dieldrin	DMDTP	MEHP	PCB 118	Xylene
		Heptachlor Epoxide	DEP	MEHHP	PCB 126	Styrene
		Hexachlorobenzene	DETP	MEOHP	PCB 156	Toluene
		Mirex	DEDTP		PCB 169	N-Hexane
		Oxychlordane	Atrazine	Parabens	Non-Dioxin-Like	2-methyl-pentane
		Trans-Nonachlor	Atrazine mercapturate	ButylparabenEthylparaben	PCB 74	3-methyl-pentane
		Endosulfan Sulfate		Methylparaben	PCB 138	
				Propylparaben	PCB 153	
					PCB 180	

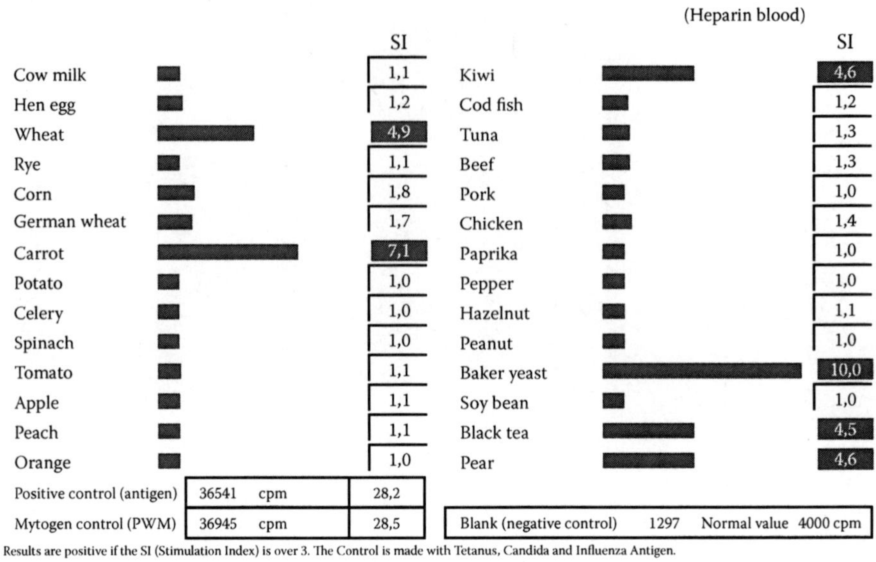

FIGURE 8.3 Lymphocyte transformation test (LTT) on food. The test shows positive reactions (delayed reaction) to wheat, carrot, kiwi, yeast, black tea, and pear.

having eaten the specific food. The symptoms can occur hours or days after ingestion. Therefore it is important to analyze the delayed type of food sensitivity.

Figure 8.3 shows the result of an LTT. Stimulation greater than threefold is a positive reaction. Patients should avoid these food for about 2 to 3 months. After this period they can try to introduce the food into their diet plan in a 4-day rotation, meaning the person can eat this food every third or fourth day.

8.5.5.1 Gluten Sensitivity Can Cause Neurological and Autoimmune Diseases

A growing problem in Europe is gluten sensitivity, which is not to be confused with celiac disease. Gluten sensitivity can lead to severe neurological reactions and brain damage. After a chemical overload, for example after the intake of antibiotics, chemicals, or drugs, the mucosal protective layer can be destroyed and the enteric immunosystem responds to undigested food. If undigested gluten contacts the surface of the mucosa a substance called zonulin (discovered by Alessio Fasano and his team at the University of Maryland School of Medicine) is liberated from enteric cells, causing the tight junctions between the mucosa cells to open, resulting in a condition known as leaky gut [22]. The consequence is increased inflammation with systemic reactions. We routinely measure zonulin and other gluten-related antibodies in patients who suffer from chronic chemical sensitivity.

8.5.6 Stool and Digestive Analysis

The new stool test uses DNA analysis to identify microbiota including anaerobes, a previously immeasurable area of the gut environment. DNA assessment is specific and accurate, avoids the pitfalls of sample transport, reports results as specific numbers, and is more sensitive than classic laboratory methods. The stool analysis detects as few as five cells per gram—a 5000-fold increase in sensitivity over microscopy for parasite detection.

8.5.6.1 Why Use Stool Analysis?

Gastrointestinal function is important for general health. The intestinal tract contains significant amounts of bacteria—some beneficial, some neutral, and some harmful. Balancing beneficial microbial flora in the gut is key to proper digestion, efficient nutrient usage, and ridding the body of waste and pathogens. Poor digestion and malabsorption can lead to immune dysfunction, nutritional insufficiencies, mental/emotional disorders, and autoimmune diseases.

8.6 THERAPY: THE DNA CONCEPT

I have summarized our therapeutic strategy as the so-called DNA concept, based on three "pillars": (1) detoxification, (2) nutritional therapy, and (3) anti-stress-management. The strategy includes

- Clean environment + Detoxification
- Orthomolecular supplements (vitamins, minerals)
- Anti-stress program: Physical therapy with sauna, massages, sport, and so on
- Individual coaching: Personal medical education of the "basics" of environmental and functional medicine

The aim is

1. Increasing detoxification capacity
2. Antiinflammatory (stop the fire in body and brain)
3. Physical and mental stability

8.7 HEPAR-TOX DETOXIFICATION

In our detoxification program, called the HEPAR-TOX therapy, we give infusions (I.V. therapy) with vitamins, minerals, and sulfur-containing chelating substances naturally present in the body such as glutathione and alpha-lipoic acid. We don't use synthetic chelators such as ethylenediaminetetraacetic acid (EDTA) or 2,3-dimercapto-1-propanesulfonic acid (DMPS).

On the basis of the aforementioned medical analysis we create an individual tailored medicine to normalize nutrient-dependent metabolic functions.

The following supplements are important tools in our detoxification therapy:

Glutathione
Alpha-lipoic acid
Coenzyme Q_{10} (ubiquinone)
Carnitin

Vitamin B complex (including niacin, pyridoxal-5-phosphate + methyl B_{12})
Vitamin D_3
Broccoli extract (sulforaphan)
Polyphenols, green tea (epigallocatechin gallate [EGCG])
Curcumin (turmeric)
Antioxidants (high-dose vitamin C, vitamin E)
Minerals and trace elements: Magnesium, zinc, selenium
Glycine

8.8 ECOLOGICAL ARCHITECTURE AND ENVIRONMENTAL MEDICINE

People who suffer from CS have to be treated in a safe environment.

In 1989, I decided to build the Institute for Environmental Diseases, applying the latest knowledge of ecological architecture in a little spa village located in Northern Hessia in the center of Germany. The lime mortar used for the plastering of the walls and the open-pored mineral pigments in the paints have the ability to filter, bind, and neutralize toxic substances. Low moisture and a short drying time as well as good insulating qualities were important criteria in selecting these building materials. We required our suppliers and contractors to provide proof that the materials delivered contained no radiation, toxic vapors, or odorous substances (low-emission products). For treatment of the wood components, terpene-free natural lacquers and specially developed waxes for sensitive patients were used.

For reduction of electromagnetic fields we have used shielded cables. To avoid oscillating fields caused by plugged-in electrical appliances, all circuits are equipped with circuit switches, which cut off electronically all current in the lines when the appliances are turned off. The circuit wiring has been laid in a star-shaped fashion.

8.8.1 New Projects in Wolfhagen—The Historic City in the Land of the Brothers Grimm

At this time we are planning, together with the city of Wolfhagen, Germany, a new ecological project: Eco-Hotel Castle Wolfhagen (Wolfhager Burg). Wolfhagen is the historic city of half-timbered houses in the fairy tale land of the Brothers Grimm. One of the Grimm Brothers, Ludwig Emil Grimm, lived in Wolfhagen. He painted numerous pictures for the famous collection of fairy tales written by his brothers Jacob and Wilhelm Grimm. The city was established in the year 1231 by Landgrave Conrad of Thuringia. Close to the Edertalsperre (Edel Valley barrier) and west of the old kurhessischen (Hessian electorate) capital of Kassel with its wealth of monuments, it is one of the most beautiful parts of the Central German Uplands region.

The number of people suffering from CS is growing constantly. These persons, who are more and more socially isolated, need safe places where they can stay, relax, enjoy holidays, or go for medical treatment.

Therefore it is a matter of urgency to build houses, hospitals, and hotels on the basis of knowledge of ecological architecture and applied environmental medicine.

FIGURE 8.4 Draft: Environmental resort, Wolfhagen Burg. Source: ANP Planning Office, Kassel, Germany.

We are privileged to get support by the local government—especially from our mayor Mr. Reinhard Schaake, who greatly supports our projects under the motto: *Environmental Technology meets Environmental Medicine.* His aim was also energy production from 100% regenerative sources by the year 2015. Wolfhagen received the "Energy Friendly Town" award from the federal government.

It took a long time but finally in February 2013 we succeeded in finishing our project plans. Figure 8.4 shows a draft of our Environmental Resort, which will be a center for cure, relaxation, and education—an island for people suffering from CS and multiple allergies. The restaurant will deliver a full spectrum of local organic and also international food that can be tolerated by people with food allergies. At the moment we are looking for partners and investors, who can be certain that they will receive the full support of our mayor and the district authorities.

ENDNOTES

1. Lichtenstein P, Holm NV, Verkasalo PK et al. 2000. Environmental and heritable factors in the causation of cancer—Analysis of cohorts of twins from Sweden, Denmark and Finland. *N Engl J Med* 343(2):78–85.
2. Rick O, Kalusche EM, Dauelsberg T, König V, Korsukéwitz C, Seifart U. 2012. Reintegrating cancer patients into the workplace. *Dtsch. Ärztebl Int* (Journal of the German Medical Association) 109(42):702–708.

3. Runow K-D. 2013. *Krebs – Eine Umweltkrankheit?* (Cancer—An environmental disease?). Munich: Südwest-Verlag.
4. Rea WJ. 1994. *Chemical sensitivity*, Vols. 1–4. Ririe, ID: Lewis Publishing.
5. Jones DS, Quinn S. 2005. *Textbook of Functional Medicine*, Gig Harbor, WA: Institute for Functional Medicine.
6. Barbara G, Wang B, Stanghellini V et al. 2007. Mast cell-dependent excitation of visceral-nociceptive sensory neurons in irritable bowel syndrome. *Gastroenterology* 132:26–37.
7. Lord RD, Bralley JA. 2008. *Laboratory evaluations for integrative and functional medicine*, 2nd ed., Duluth, GA: Metametrix Institute.
8. Galland L, Lafferty H. 2008. *Gastrointestinal dysfunction: Connections to chronic disease*. A Functional Medicine Monograph. Gig Harbour, WA: The Institute for Functional Medicine.
9. Runow K-D. 2009. *Wenn Gifte auf die Nerven gehen*, 2nd ed. (If toxic chemicals attack the nervous system: How we can protect our brain and nerves by detoxification). Munich: Südwest-Verlag/Random House Verlagsgruppe.
10. Runow K-D. 2011. *Der Darm denkt mit*, 6th ed. (The gut-brain connection; How bacteria, yeast and allergies attack the nervous system). Munich: Südwest-Verlag/Random House Verlagsgruppe.
11. Alschuler LN, Gazella KA. 2010. *The definitive guide to cancer: An integrative approach to prevention, treatment, and healing*. New York: Crown Publishing Group.
12. Metametrix Institute. 2009. Documented limitations of culture based stool assessment, www.metametrix.com.
13. Clarridge JE, 3rd. 2004. Impact of 16S rRNA gene sequence analysis for identification of bacteria on clinical microbiology and infectious diseases. *Clin Microbiol Rev* 17(4):840–862.
14. Ghosh S, Debnath A, Sil A, De S, Chattopadhyay DJ, Das P. 2000. PCR detection of *Giardia lamblia* in stool: Targeting intergenic spacer region of multicopy rRNA gene. *Mol Cell Probes* 14(3):181–189.
15. Langendijk PS, Schut F, Jansen GJ et al. 1995. Quantitative fluorescence in situ hybridization of *Bifidobacterium* spp. with genus-specific 16S rRNA-targeted probes and its application in fecal samples. *Appl Environ Microbiol* 61(8):3069–3075.
16. Welling GW, Elfferich P, Raangs GC, Wildeboer-Veloo AC, Jansen GJ, Degener JE. 1997. 16S ribosomal RNA-targeted oligonucleotide probes for monitoring of intestinal tract bacteria. *Scand J Gastroenterol Suppl* 222:17–19.
17. Delgado S, Suarez A, Mayo B. 2006. Identification of dominant bacteria in feces and colonic mucosa from healthy Spanish adults by culturing and by 16S rDNA sequence analysis. *Dig Dis Sci* 51(4):744–751.
18. Lakind JS, Naiman DQ. 2010. Daily intake of bisphenol A and potential sources of exposure: 2005–2006 National Health and Nutrition Examination Survey. *J Exp Sci Environ Epidemiol* 21(3):272–279.
19. Braun JM, Kalkbrenner AE, Calafat AM et al. 2011. Variability and predictors of urinary bisphenol A concentrations during pregnancy. *Environ Health Perspect* 119(1):131–137.
20. Metametrix Clinical Laboratory. 2011. *Bisphenol A profile guide*. Duluth, GA: Metametrix Institute, www.metametrix.com.
21. Genova Diagnostics. 2012. *Toxic effects: CORE. Interpretive guide*. Asheville, NC: Genova Diagnostics, www.gdx.net.
22. Fasano A, Flaherty S. 2014. *Gluten freedom*. New York: Wiley General Trade.

9 Emission Rate of Chemical Compounds in Building Products and Materials

Shin-ichi Tanabe, PhD, Professor[1] *and Chairman*[2]
[1]Department of Architecture, Waseda University
[2]ISO/TC146/SC6 for Indoor Air

9.1 INTRODUCTION

Residential buildings in Japan use air-tightened building envelops in order to conserve energy. Without an adequate ventilation system, however, there is a reduction in outdoor air intake. In addition, building today make use of modern building products to reduce costs and simplify construction procedures. These products contain various chemical substances that cause what is known as sick-building syndrome (SBS) in Japan [1].

It is very important to prevent SBS, which is caused by chemical off-gasing, by using low-emission materials and maintaining an adequate ventilation rate. To achieve a low chemical concentration in the air, low chemical emission materials can be employed. The Japanese government enforced a new building code in July 2003. The code restricts areas using formaldehyde-emitting materials, prohibits the use of chlorpyrifos, and mandates a minimum ventilation rate using a mechanical system. This code describes measurements and evaluation methods for chemical emission rate. To avoid chemical pollution, increasing the ventilation rate is beneficial, using low-emission materials is a higher priority. In addition to the building materials, furniture, electric appliances, and cleaning products all emit chemicals.

9.2 CATEGORIES OF CHEMICAL SUBSTANCES

Table 9.1 lists the categories of indoor chemical substances. Chemical substances are designated by their boiling points.

Several methods can be employed to measure chemical emission rates, including the small-chamber, large-chamber, Field and Laboratory Emission Cell (FLEC), emission cell, desiccator, and microchamber. Both the chamber and FLEC methods are referred to as dynamic headspace methods in which fresh air is pumped into a chamber, and

TABLE 9.1
Indoor Chemical Substances

Category	Description	Abbreviation	Boiling Point Range (°C)[a]	Sampling Methods Typically Used in Field Studies
1	Very volatile (gaseous) organic compounds	VVOC	<0 to 50–100	Batch sampling; adsorption on charcoal
2	Volatile organic compounds	VOC	50–100 to 240–260	Adsorption on Tenax, carbon molecular black, or charcoal
3	Semivolatile organic compounds	SVOC	240–260 to 380–400	Adsorption on polyurethane foam or XAD-2
4	Organic compounds associated with particulate matter or particulate organic matter	POM	>380	Collection on filters

[a] Polar compounds appear at the higher end of the range.

the outlet air concentration is measured from another chamber. The chamber method allows the measurement of both VOCs and VVOCs. However, SVOCs are difficult to measure with a normal chamber, requiring the use of the microchamber method. In addition, there are methods that do not use ventilation. Formaldehyde is measured using a 20-L glass desiccator, and it is absorbed by distilled water in a dish. The concentration in the water is then analyzed by the absorption spectrum method.

The emission chamber method is also a dynamic headspace method that simulates a normal room, allowing for high-quality measurements. The 20-L chamber method is widely used in Japan.

9.3 BUILDING PRODUCTS AND MATERIALS

Typical building products and materials include wooden materials, wallpaper, thermal insulation, adhesives, and furniture. Unpainted glass and metal materials usually have low-emission characteristics. The majority of indoor emissions generally come from interior surfaces such as ceilings, walls, and floors. However, adhesives are often used on the contact surfaces of building products, and these may emit relatively high quantities of VOCs. In addition, VOCs can be emitted through the surface via diffusion.

9.4 EMISSION TESTS USING A CHAMBER

9.4.1 Small-Chamber Method

The Japanese Industrial Standard (JIS) A 1901 [2] describes the small-chamber method of measurement for a chamber volume of 20 to 1000 L. Small-sized chambers are widely

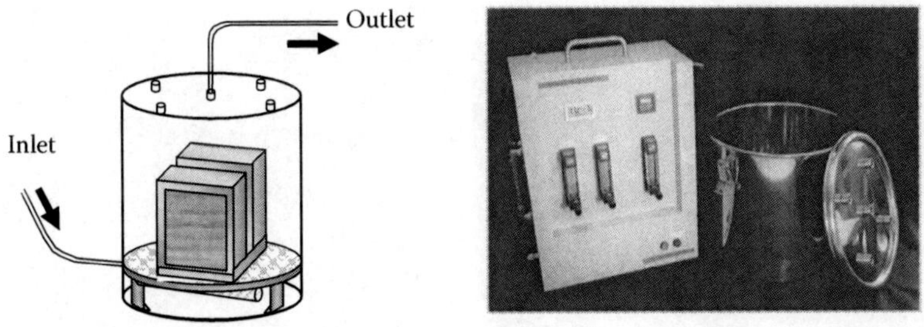

FIGURE 9.1 20-L small-chamber system (ADPAC).

used in Japan, the United States, and Europe [3,4]. In Japan, a 20-L chamber is commonly used (Figure 9.1), which consists of an SUS304 chamber, air purifier device, air control unit, and mixing chamber. The SUS304 chamber is installed in a constant-temperature chamber. The building materials are prepared to a size of 165 mm × 165 mm in a sealed box to prevent emission. Purified air is supplied at 167 mL/min (air change rate of 0.5 1/h). The outlet air from the chamber is collected with a Tenax® TA sorbent tube. The measurements are taken 1, 3, and 7 days after the test starts. Gas chromatography (GC)/GM is used for chemical analysis to measure the amount of aldehydes and VOCs. A sorbent tube is typically analyzed with thermal desorption via gas chromatography–mass spectrometry (GC–MS), a gas chromatography–flame ionization detector (GC–FID), or high-performance liquid chromatography (HPLC). This test determines the emission factor, which is calculated by the outlet concentration of the emission chamber, ventilation rate, and surface area of the VOCs, formaldehyde, and other carbonyl compounds released from building products.

The emission rate of the target chemical is described in Equation 9.1.

$$\mathrm{EF} = \frac{(C_t - C_{\mathrm{tb},t}) \times Q}{S} \tag{9.1}$$

where

EF = Emission rate per unit surface [μg/(m$^2 \cdot$h)]
C_t = Chamber concentration at time t [μg/m^3]
$C_{\mathrm{tb},t}$ = Travel blank concentration at time t [μg/m^3]
Q = Ventilation rate [m^3/h]
S = Surface area of material [m^2]

9.4.2 Large-Chamber Method

Large building products, furniture, "System Kitchens," and electric appliances can be measured without removal of material. In Japan, a large chamber is defined as one with a volume between 1 m^3 and 80 m^3. The principle of measurement is the same as that of the small-chamber method, with the emission rate being obtained by Equation 9.1 [5,6]. The results are often described by the measurement [μg/(unit $\cdot$ h)] instead of the amount per surface area per unit time. Figure 9.2 shows an example of a large chamber.

(a) (b)

FIGURE 9.2 Large chamber. (a) Outside view of chamber. (b) Emission test (with MDF).

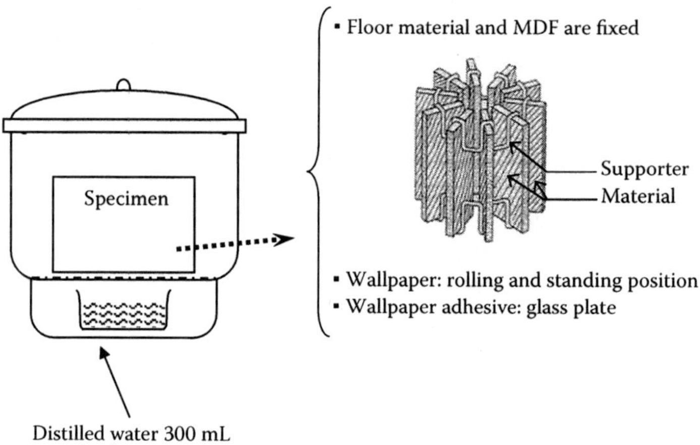

FIGURE 9.3 20-L glass desiccator method.

9.4.3 Desiccator Method

Formaldehyde emission from wooden products, wallpaper, and adhesives used for wallpaper is often measured using the desiccator method; this is much simpler than the small-chamber method. Both Japanese Agricultural Standards (JAS) and JIS A 1460 [7] are widely applied. Building materials used indoors require display of a formaldehyde emission label. Figure 9.3 shows the desiccator method. An acrylic desiccator is sometimes used for special wooden materials.

9.4.4 Passive Method

An Advanced Diffusive Sampling Emission Cell (ADSEC) is a device containing a diffusive sampler for measuring VOC emission rates from building materials. An aluminum basement ensures the airtight performance of the ADSEC. Figure 9.4 shows the configuration of the ADSEC. It measures building materials cut to a size

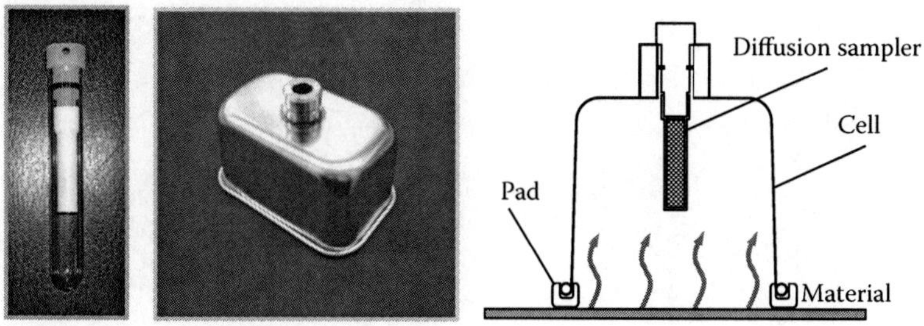

FIGURE 9.4 Emission small cell.

of 165 mm × 165 mm, which is the regulation size in JIS A 1903 [8]. An ADSEC is made of an SUS304 stainless steel cell (length: 59 mm, width: 97.5 mm, height: 56 mm, volume: 300 mL). The emission rate is calculated by dividing the collected amount of VOCs from the test area (0.00494 m^2) by the sampling time (24 h), as shown in Equation 9.2. VOC–SD was often used as the diffusive sampler, which is also shown in Figure 9.4. The VOC–SD was analyzed with GC–MS after solvent desorption and thermal desorption.

$$J_A = \frac{M}{S_{ad} \times t} \tag{9.2}$$

where

J_A = Emission rate (passive method) [μg/($m^2 \cdot h$)]
M = Total amount of collected substances [μg]
S_{ad} = Covered surface area [m^2]
t = Test duration [h]

9.4.5 Microchamber Method for SVOCs

The principle of the test is to determine the area-specific emission rates of SVOCs emitted from the surface of a product test specimen [9]. Although SVOCs are emitted in the microchamber, the majority of these emissions are adsorbed in the chamber at temperatures of 40°C or below. Therefore, in this test, the area-specific emission rate of SVOCs from a building material is determined from the mass collected in the first and second steps. The outcome of the test is the mean rate of emission of SVOCs from the product over a 24-h period. For specific purposes, the emission rate over a different period of time could be determined using the same procedure, but with a different duration of the first step. Figure 9.5 shows a microchamber for SVOC measurement.

The mass collected in the control test (field blank) with a clean microchamber must be less than 10% as compared with the total mass of the target SVOC. The control test confirms the recovery rate. This method requires no direct contact between the test specimen and the inner walls of the microchamber, as shown in Figure 9.6. During the first step of the test, the concentration measurements are carried out at

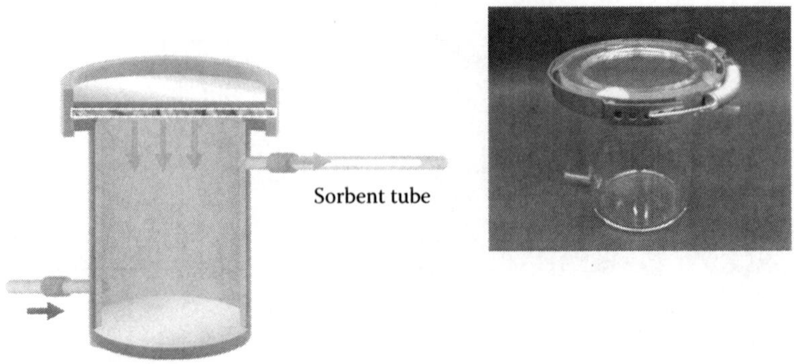

FIGURE 9.5 Microchamber for SVOC measurement.

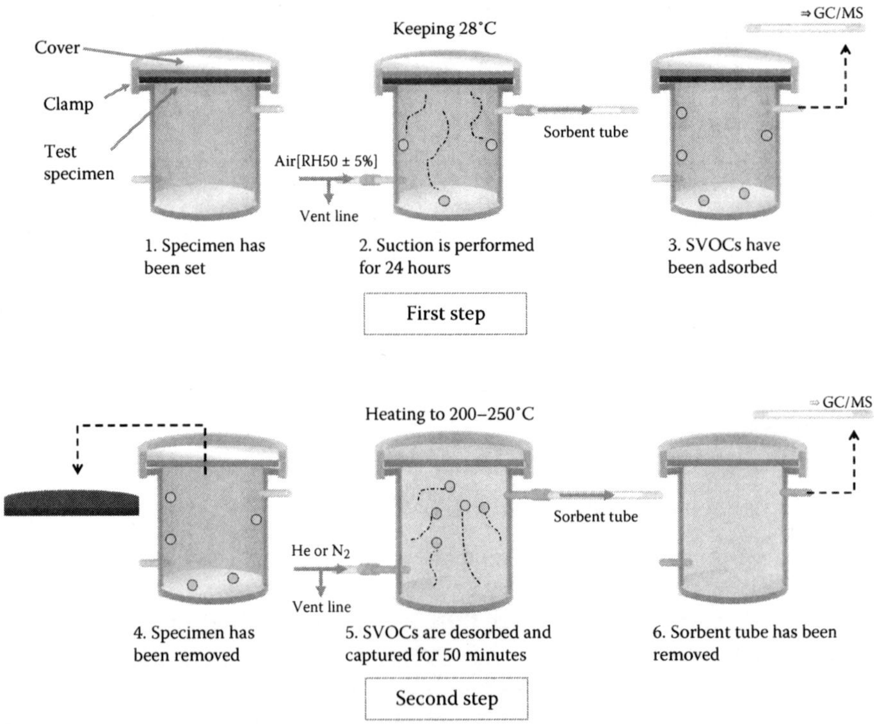

FIGURE 9.6 Two-step measurement.

predefined sampling times, which is typically every 24 h. Depending on the objective of the test, it can be appropriate to sample the air at additional times. The air-sampling duration for the concentration measurements depends on the analytical methods. Humidified clean air (50% RH) is supplied during the air sampling in the emission test.

During the desorption test (the second step of the test), the test specimen is removed from the microchamber, and the microchamber is heated. A different conditioned sorbent tube is used for this part of the test. Inert gas is supplied and then heated to 200°C or above.

The sampling and ventilation of the microchamber must be started simultaneously. The sampling time is 40 min until the desorption is complete.

Calculate the area-specific emission rate of the test specimen using Equation 9.3:

$$\mathrm{EF} = \left(\frac{m_\mathrm{n}}{Q \cdot t} \cdot \frac{Q}{S}\right) + \frac{m_\mathrm{d}}{S \cdot t} = \frac{m_\mathrm{n} + m_\mathrm{d}}{S \cdot t} \tag{9.3}$$

EF = Emission rate [μg/(m^2 · h)]
m_n = Collected amount during the first test step [μg]
m_d = Collection amount during second test step [μg]
Q = Chamber ventilation rate [m^3/h]
S = Exposed surface area of specimen [m^2]
t = Total time during first test step (typically 24 h) [h]

9.5 LABELING

The new requirements of the amended Building Standard Law on SBS issues are effective only for buildings with construction that began on or after July 1, 2003. The building materials that are subject to the law are listed by the Ministry of Land, Infrastructure, Transport and Tourism, 2002 [10]. Thus any built-in furniture, kitchen cabinets, and other products designed as interior building materials that require chemical, notification are also subject to the law. When using interior finishing materials that require notification, these materials must be classified according to their class (grade) of formaldehyde emission. Materials that do not require notification, such as stones and metals, can be used without restriction for interior finishing materials in habitable rooms, because these materials emit low level of formaldehyde.

9.5.1 FORMALDEHYDE

Table 9.2 lists the new formaldehyde emission standards from wooden building materials, as determined according to the desiccator method by the Ministry of Agriculture, Forestry and Fisheries. It provides the grading that ensures buildings are compatible with the small chamber method and JIS A 1901.

The symbols for grading formaldehyde emissions are as follows: F☆☆☆☆ denotes building materials that are not restricted and these emit low levels of formaldehyde, and F☆☆☆ through F☆☆ denote restrictive use of materials.

Building materials requiring regulation include coating materials, such as insulation materials (including plywood, wood flooring, particle board, medium density fiberboard [MDF], etc.), wallpaper, formaldehyde, glue, paint, and wooden finishing.

TABLE 9.2
Labeling Formaldehyde Emissions in Japan

Symbol (Category)	Building Code			Limitation for Indoor Use
	Chamber Method	Desiccator Method		
		Average	Max	
F☆☆☆☆	Below 5 μg/m²·h	0.3 mg/L	0.4 mg/L	No restriction
F☆☆☆	5–20 μg/m²·h	0.5 mg/L	0.7 mg/L	Limited area for indoor use
F☆☆	20–120 μg/m²·h	1.5 mg/L	2.1 mg/L	
F☆	Above 120 μg/m²·h	5.0 mg/L	7.0 mg/L	Use prohibited indoors

9.5.2 VOCs

While the Ministry of Health and Welfare published guideline values of 13 chemical substances of emission rates from building products are not currently regulated for chemicals other than formaldehyde.

9.5.2.1 For VOC Standard

In response to the establishment of the small-chamber method by JIS A 1901, requirements have been determined for the dissipation of VOC emissions from building materials [11]. Manufacturers of building products can then initiate the quality control of their materials using this common method.

Table 9.3 lists the voluntary-based VOC emission rate limit values. Unfortunately, TVOC has not been defined after extended discussions with industry in Japan. It is assumed that the measurement results on the seventh day by the chamber method provide the TVOC emission. Furthermore, in the calculation of the emission rate reference value, we assumed that the conditions were the same as the basis of the SHS building measurement technical standards of the Building Standard Law: a ventilation rate of 0.5 1/h and a temperature of 28°C.

TABLE 9.3
Target VOC and Emission Rate

VOC	Symbol	Limit Emission Rate [μg/(m²·h)]	Guideline Value [μg/m³]
Toluene	T	38	260
Xylene	X	120	870
Ethylbenzene	E	550	3800
Styrene	S	32	220

Source: Japan Testing Center for Construction Materials. 2008. Standard for VOCs emission from building materials. (in Japanese)

ENDNOTES

1. Tanabe SI. 1998. *Chemical pollution*. Tokyo: Kodansha. (in Japanese)
2. Japanese Industrial Standards Committee, JIS A 1901. 2015. Determination of the emission of volatile organic compounds and aldehydes for building products: Small chamber method.
3. ISO, ISO 16000-9. 2006. Determination of the emission of volatile organic compounds from building products and furnishing: Emission test chamber method.
4. Wolkoff P, Clausen PA, Nielsen PA, Gustafsson H, Jonsson B, Rasmusen E. 1991. *Field and laboratory emission cell: Proc. of FLEC-IAQ 91 healthy buildings*, pp. 160–165.
5. Japanese Industrial Standards Committee, JIS A 1911. 2015. Determination of the emission of formaldehyde for building materials and building related products: Large chamber method.
6. Japanese Industrial Standards Committee, JIS A 1912. 2015. Determination of the emission of volatile organic compounds and aldehydes without formaldehyde for building materials and building related products: Large chamber method.
7. Japanese Industrial Standards Committee, JIS A 1460. 2015. Building boards determination of formaldehyde emission: Desiccator method.
8. Japanese Industrial Standards Committee, JIS A 1903. 2015. Determination of the emission of volatile organic compounds (VOC) for building products: Passive method.
9. Japanese Industrial Standards Committee, JIS A 1904. 2015. Determination of the emission of semi volatile organic compounds by building products: Micro chamber method.
10. Ministry of Land, Transport and Infrastructure. 2003. *Manual for sick house syndrome, Building code and housing performance indication system*. (in Japanese)
11. Japan Testing Center for Construction Materials. 2008. Standard for VOCs emission from building materials. (in Japanese)

10 Ventilation Strategies for Each Kind of Building and Statutory Regulations

Haruki Osawa, PhD, Senior Researcher
Department of Environmental Health,
National Institute of Public Health (NIPH)
Masaki Tajima, PhD, Associate Professor
School of System Engineering, Kochi University of Technology

Regardless of their ownership or scale of use, buildings are important components as social capital that supports the lives of people and industrial activity. In Japan, the Building Standard Law was enacted to safeguard the lives, health, and property of people by providing minimum standards concerning the site, construction, and use of buildings. Under the Building Standard Law, a strict regulation was designed to prevent construction without permission (i.e., building certification) following an onsite inspection by the building official of the local government or the designated institute. Furthermore, in accordance with an amendment of the Buildings Standards Law (addition of Article 28:2: "Sanitary measures to prevent adverse effects of chemical substance emissions in habitable rooms" and promulgation of related governmental ordinances and notices), beginning July 2003 use of building materials that contain chlorpyrifos has been forbidden, while the use of building materials containing formaldehyde has been restricted. Moreover, it became mandatory to install ventilation equipment in all habitable rooms, in other words, in all building spaces that people constantly use or stay in.

In addition to the regulatory measures of the Building Standard Law, two other laws ensure standards of indoor air quality through the promotion of multifaceted countermeasures, which also combine some inductive measures. First, the Housing Quality Assurance Act provides guidelines for evaluation, rule adjustment, and enhancement of air quality performance indicator systems. Second, the Act on Maintenance of Sanitation in Buildings requires an approval of initial air quality in buildings of public interest.

After reviewing the technical background and history of indoor environmental issues in Japan, the outlines and relevant passages of the Building Standard Law, Housing Quality Assurance Act, and Law for Maintenance of Sanitation in Buildings referring to ventilation and air quality are described. Then, the current revisions in these laws are explained.

10.1 HISTORY AND BACKGROUND OF INDOOR AIR POLLUTION MEASURES IN JAPAN

10.1.1 Before the Dawn of Measures against Sick Houses

Indoor air pollution is caused by stagnation of matter in buildings that is harmful to human health. Consequently, disregarding the effects of absorption or decomposition, there are only two basic measures: to halt or mitigate the inflow and to accelerate the outflow of pollutants. However, in reality the procedure is not so simple at actual sites.

Formerly, traditional environmental engineering had recognized atmospheric pollution, exhaust air from combustion, or the human body as major source of indoor air pollution; the Building Standard Law therefore had recommended a mechanical or natural ventilation scheme.

Modern architecture employs many chemicals and chemical products to carry out bondability, plasticity, insect-proofing, preservation, mold prevention, fireproofing, and so on. Today, such chemicals are essential to design or build efficient and comfortable buildings at affordable cost.

While extreme reduction of the ventilation rate from the 1980s was aimed at energy conservation, it led to health effects called the "sick-building syndrome" (SBS) in developed countries. In Japanese buildings, levels of harmful particulates or gas increased and were detected by researchers in the 1990s while smoking was declining. However, Japanese buildings escaped serious damage thanks to the guidelines on the concentration of carbon dioxide in the law on building sanitation and the Building Standard Law. On the other hand, inhabitants of Japanese dwellings suffered health problems due to SBS, which was not covered by protective regulations and the buildings were easily air-tightened. In Japanese dwellings, considering the thermal efficiency and prevention of concealed condensation in structures, use of vapor and air barriers was strongly recommended after the oil crisis. This change of layer design led to a rapid upgrade of air-tightness and reduction of natural ventilation.

In addition, social changes in Japan, such as a decline in family size per household, an increase of two-income families, a change of daily schedules, degradation of the outdoor environment, and so forth may have reduce the custom of airing. Many researchers clarified in detail the synergic effect of the reduction of air change and the increase of chemical emissions in dwellings.

However, in those days, there were many problems standing in the way of rational planning and implementation of concrete countermeasures, as follows.

1. Undeveloped criteria for evaluation and measurement (causing difficulty in setting a rational standard)
2. Inadequate techniques for measurement and analysis (causing difficulty in grasping concentration and emission rate)
3. Unfounded mechanism of pollution propagation (leading to a lack of practical evidence for contamination moving)

Adding to these main obstacles, other problems extending to architecture, chemistry, mechanical engineering, and medical science were mounting.

To deal with this difficult situation, the Ministry of Land, Infrastructure, Transport and Tourism, Ministry of Construction (MLIT; name used at that time) started working on knowledge accumulation through collaboration with the Ministry of International Trade and Industry, Forestry Agency, Ministry of Health and Welfare, Forestry Agency (names used at the time), relevant organizations, and academia. Then, by organizing the Research Consortium for Healthy Housing from 1996 and Research Committee on measures for Indoor Air from 2000, studies made progress and data gradually accumulated.

10.1.2 Design Assumption and Strategy of Measures against the Indoor Air Pollution Problem

Prior to enforcing such strong measures as legal regulations, it is essential to provide effective countermeasures based on an accurate overview of the emission and propagation mechanism of contaminants and their hazards. Another crucial obligation during such administrative procedures is to explain the consequences and the necessity of such measures to the citizens, who will end up bearing the cost and disadvantages. However, in addition to the inherent toxicity of the aforementioned substances, damages arising from indoor air pollution tend to depend also on other factors, such as the concentration of the substance, exposure time, and the sensitivity of the human body to various substances. In other words, even if there is the same amount of a poisonous substance in the air, depending on its concentration and duration there are cases where no health issues may arise. Thus, risk assessment and formation of consistent countermeasures remain difficult.

When designing the formaldehyde regulation as provided in the Building Standard Law, and when considering the causes of health hazards to residents, four variables were taken into account as the main factors that affect the concentration of indoor air pollution: (1) Emission, (2) Exhaust, (3) Adsorption and Desorption, and (4) Room Volume. Of the four, Adsorption and Desorption and Room Volume were taken as independent variables during the formulation of regulatory measures concerning Emission and Exhaust. Also, because various technological developments regarding Adsorption and Desorption are anticipated, it is considered that the implementation of a ministerial approval system might boost those technological development effects. Considering Room Volume, standards on the number of ventilation equipment are eased in cases of rooms with high ceilings because of the higher indoor air volume.

However, in this regard, the standards depend on the emission and propagation mechanisms. If the physical property or the toxicity were to change, different relations would have to be taken into consideration.

As mentioned previously, indoor pollution concentration is based mainly on the rates of pollution emitted and exhausted. Furthermore, the rates of emission depend on the diffusion speed/area of the emission source and air inflow from adjacent spaces. The exhaust rate is affected by the ventilation rate and the collection efficiency.

This much is clear: Various approaches exist when formulating countermeasures for reducing harmful effects of formaldehyde, but only two have been enforced by the Building Standard Law. Restrictions on the use of certain building materials aim at decreasing the amount of emissions, while the mandatory installation of ventilation equipment aims at ensuring a steady exhaust flow, thus securing satisfactory indoor air quality.

Based on these assumptions, these two strategies have been adopted as the basis of regulations in the Building Standard Law, to constrain air inflow from adjacent concealed spaces (e.g., from attic spaces) by using supplementary adsorption measures (e.g., panels, painting, or coating) to reduce overall emissions. Because the law considers air inflow from within the building framework (such as from attic cavities or concealed spaces inside of walls), and has regulations that specifically refer to ant repellants, it makes it unique to Japan without precedents elsewhere in the world.

Despite having a two-sided strategy that addresses both the emission source and ventilation part of the problem, there have been counterarguments asking: "Can't you deal with just one side of the issue?" However, even contemporary medical and chemical technology cannot fully specify the toxicity of all building materials, let alone find perfect substitutes for them or completely exclude problematic building materials.

Also, although people are still unaware of possible harm from long-term exposure to low concentrations of indoor air pollution, one cannot say it is realistic to expect that potential risks will be avoided only through an increase in ventilation. Such measures could impair indoor thermal environment or even cause further burden to air-conditioning devices. On carefully considering the impacts of alternative substances, building materials, and functions, there is a need for a comprehensive and balanced discussion regarding the Building Standard Law, its risks, effectiveness, and reliability as a system. The Ministry of Land, Infrastructure and Transfer regulations are summarized in Figure 10.1.

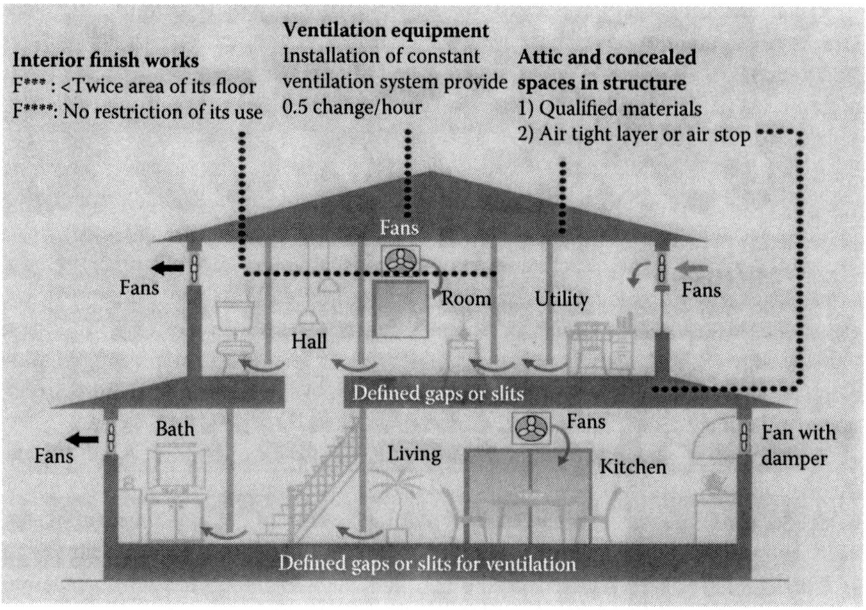

FIGURE 10.1 Summary of housing regulations for by the Ministry of Land, Infrastructure, and Transfer.

10.2 STATUTORY REGULATIONS FOR INDOOR AIR QUALITY

10.2.1 Technical Standards in the Amended Building Standard Law

The following is a summary of the "Partial Amendment to the Building Standard Law (Act No. 85/2002)," in particular the article referring to "Sanitary measures to prevent adverse effects of chemical substance emissions in habitable rooms" enforced on July 1, 2003:

1. Chemical substances to be regulated by the order are chlorpyrifos and formaldehyde.
2. Regulations are included regarding the use of building materials containing chlorpyrifos.
3. The use of building materials containing chlorpyrifos is prohibited. (However, building materials that have been parts of a building for more than 5 years are exempted from this regulation.)

10.2.2 Regulation Regarding the Use of Building Materials Containing Formaldehyde and Installation of Ventilation

10.2.2.1 Restrictions on Interior Finishing

When completing interior work on habitable rooms (including parts of the building, such as hallways, where ventilation will be shared between that space and the habitable room through the installation of a permanent opening, such as door undercuts), the usable area for building materials that contain formaldehyde should be decided based on the emission rate of the material under the specified conditions. Building materials are classified as prohibited (type 1) or restricted (type 2, type 3) based on the habitable room type and the ventilation rates. Regarding the restricted materials, the usable area for such materials is calculated using a specified formula designated to adhere to the concentration in the guideline.

10.2.2.2 Mandatory Installation of Ventilation

In principle, even when formaldehyde-emitting building materials are not used, to deal with emissions from furniture, installation of ventilation equipment is required for all buildings. The required effective ventilation air volume (or the amount of effective conversion in cases in which the equipment has an air purifying function) is based on the type of ventilation equipment.

Still, exemptions from these requirements apply in cases in which conditions related to air tightness (more than 15 cm^2/m^2 of Equivalent Leakage Area [ELA], the amount of equivalent opening area per floor area of building) and structural requirements (buildings of "shinkabe" [traditional timber structure with plaster wall] construction in which no plywood or similar panel-like building material is used for exterior walls, ceilings, and floors) are fulfilled, and the buildings are deemed to have the air exchange rate of at least 0.5 times/h. Also the required amount of ventilation equipment is reduced for spaces with high ceilings. Moreover, spaces in which people are expected to be regularly active and that have been deemed able to maintain the formaldehyde concentration levels below 0.1 mg/m^3 throughout the year can receive a certification from the Minister of Land, Infrastructure and Transport. Such habitable rooms are also exempted from the regulations on building materials.

10.2.2.3 Restrictions Concerning Attic and Adjacent Concealed Spaces

Where mechanical ventilation equipment or a centrally controlled air conditioning unit is installed in place, to restrain formaldehyde inflow to habitable rooms from attics or other adjacent concealed spaces, any of the following measures should be taken:

1. For backing materials, thermal insulation materials, and panel-like materials, the following building materials must not be used:
   a. Type 1 formaldehyde-emitting building materials
   b. Type 2 formaldehyde-emitting building materials
   c. Building materials approved by the Minister of Land, Infrastructure and Transport under the provisions of Article 20-5, paragraph 2 of the Order (building materials deemed to be equivalent to type 2 formaldehyde-emitting building materials)
2. Install an air-tight layer to stop air circulation and formaldehyde inflow into habitable rooms.
3. Install ventilation equipment that would keep air pressure of habitable rooms higher than the air pressure of adjacent spaces, thus inhibiting inflow of formaldehyde into habitable rooms by air pressure regulation.

10.2.3 Technical Standards in the Housing Quality Assurance Act

The objective of the Housing Quality Assurance Act is to promote fair purchase and use of housing to consumers through the establishment of objective standards for performance of housing and systems of evaluation based thereon.

In contrast to the previously mentioned Building Standard Law, which was enacted to safeguard the lives, health, and property of people by providing minimum standards concerning the site, construction, and use of buildings, the Housing Quality Assurance Act is an inductive system that provides information and protects consumers (including systems of liability provision for warranties against defects).

Alongside provisions for structural safety, fire safety, durability, low energy consumption, and considerations for the elderly, provisions concerning performance evaluation and labeling of air environments have been given under the following headings: Formaldehyde Countermeasures Classification, Ventilation Measures, and Concentration of Chemical Substances in Indoor Air. Still, the provision on Concentration of Chemical Substances in Indoor Air stipulates indoor concentration measurement and analysis methods of target substances (formaldehyde, toluene, xylene, ethylbenzene, and styrene) after completion of construction. Because this provision is based on numerical data it is omitted here.

10.2.4 Technical Standards of the Act on Maintenance of Sanitation in Buildings

The Act on Maintenance of Sanitation in Buildings protects environmental health from the perspective of maintenance and operation, and was enacted in 1970 to maintain and promote public sanitation (Figure 10.2). Therefore the subject of this legislation is limited to the use and operation aspect of buildings intended for general public use.

Also, because it is not a policy that directly intervenes in issues of health and environments of individuals, in principle, it does not directly intervene either on issues of design and/or execution at the time of construction or on waste disposal at the time of demolition. However, at the time of building operation, there are some design and operation method issues that are difficult to cope with, so it is considered that the health care center should guide operations.

From April 2003, to advance the coordination between the Act on Maintenance of Sanitation in Buildings and the revisions of the Building Standard Law, formaldehyde concentration measurement (certification) has been mandatory during the first summer after building completion. For some time now, sanitation management standards have been established, and measurements of air environments have been implemented once, or more, every 2 months (excluding formaldehyde measurements). As a result, the required maintenance of air conditioning equipment does not permit reduction of the ventilation amount. This regulation is an important factor when explaining why Japan was able to avoid the severe SBS problem that arose in other countries. The formaldehyde measurement method and management standards are modeled after the Building Standard Law. The guidelines for sanitation management are given in Table 10.1.

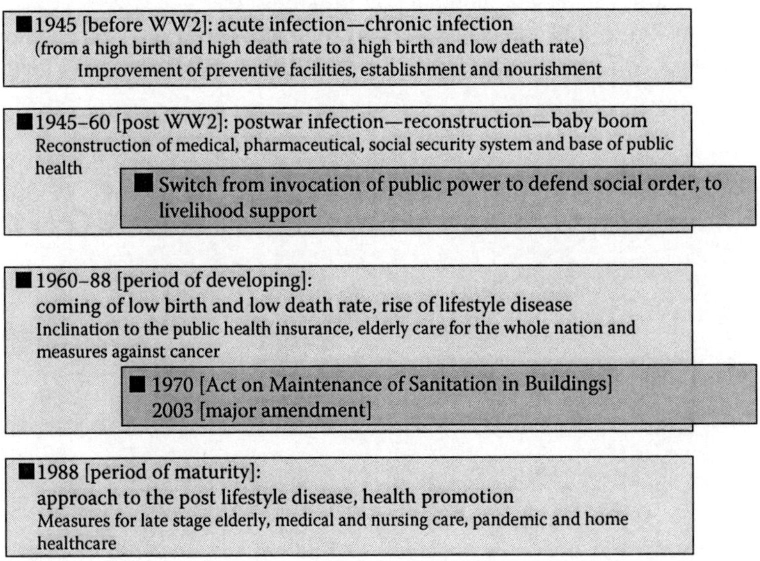

FIGURE 10.2 Background and current status of building sanitation in Japan.

TABLE 10.1
Guidelines for Sanitation Management

Indoor Environment	Standard Level
Particulate matter	<0.15 mg/m^3
CO	<10 ppm
CO_2	<1000 ppm
Temperature	• 17°C < t < 28°C • When indoor temperature is under the ambient temperature, the difference should be moderate
Relative humidity	40% < RH < 70%
Air velocity	<0.5 m/s
Formaldehyde	<0.1 mg/m^3 (<0.08 ppm)

As previously shown, if a centrally specified air conditioning system is equipped, the building passes checks, such as the emission source regulations. Because of this, the actual performance is guaranteed by this regulation. However, in research from recent years, one can come across cases in which the results of the air environment measurements were incompatible with the sanitation management standards.

Figure 10.3 shows the aggregated results (summarized by the Ministry of Health, Labour and Welfare) of on-site inspections performed from 1996 to 2008 at specified designated buildings. To be specific, Figure 10.3 shows the chronologic overview of changes in the ratio of buildings that were deemed inadequate (hereafter referred to as inadequacy rate) following the periodic air environment measurements mandated by the Act on Maintenance of Sanitation in Buildings.

As seen in Figure 10.3, the inadequacy rate of relative humidity, temperature, and carbon dioxide concentration is higher than that of other factors. In particular, the

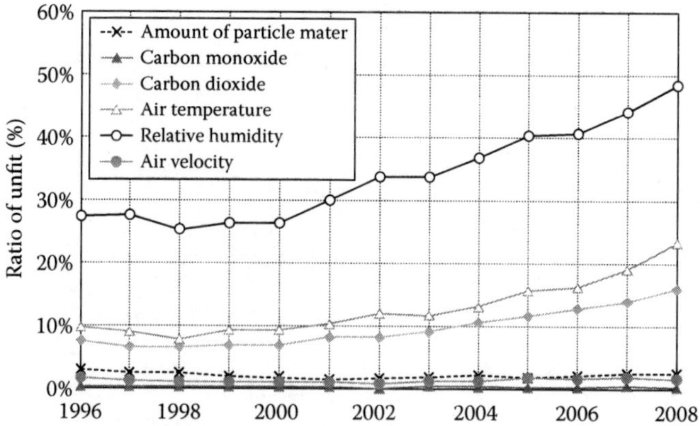

FIGURE 10.3 Trend of unfit ratio in specified buildings by environmental factors from a national database.

inadequacy rate of relative humidity has been remarkable, with the rate exceeding 40% in recent years.

10.2.5 Indoor Air Pollution Following the Amendment of the Building Standard Law in 2003

After the survey in 2000, to find the state and changes in indoor air pollution levels due to volatile organic compounds (VOCs) in houses in Japan and to confirm the effect of the amendment to the Building Standard Law, measurements of indoor VOC concentrations and investigations on the actual conditions in the residential environment were continued. These surveys covered a total of more than 10,000 newly built houses over 6 years (from 2000 to 2005) under the leadership of MLIT. The VOCs initially measured were formaldehyde, toluene, xylene, and ethylbenzene, followed by the subsequent inclusion of styrene and acetaldehyde.

Figure 10.4 shows the change in the mean concentration of formaldehyde during the period of the survey. The mean concentration in 2000 was similar to the guideline level (0.08 ppm) established by the Ministry of Health, Labour and Welfare of Japan, while the values measured in 27% of the houses exceeded this guideline value. Similarly, the mean concentration of toluene in 2000 also exceeded the guideline level (0.07 ppm) in 12% of the houses investigated; however, it showed a tendency to decrease. While the reduction in the levels of these two pollutants was smaller in 2003 when the amended Building Standard Law was enforced, levels of both compounds decreased considerably after 2004, reaching the same reduction curves as those before 2003.

As shown in Figure 10.5, these concentrations of indoor pollutants rise every summer depending on temperature change.

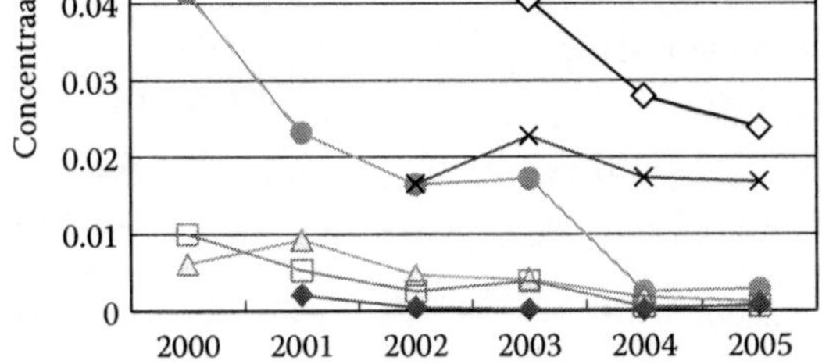

FIGURE 10.4 Trend of concentrations of dwellings following the amendment to the Building Standards Law.

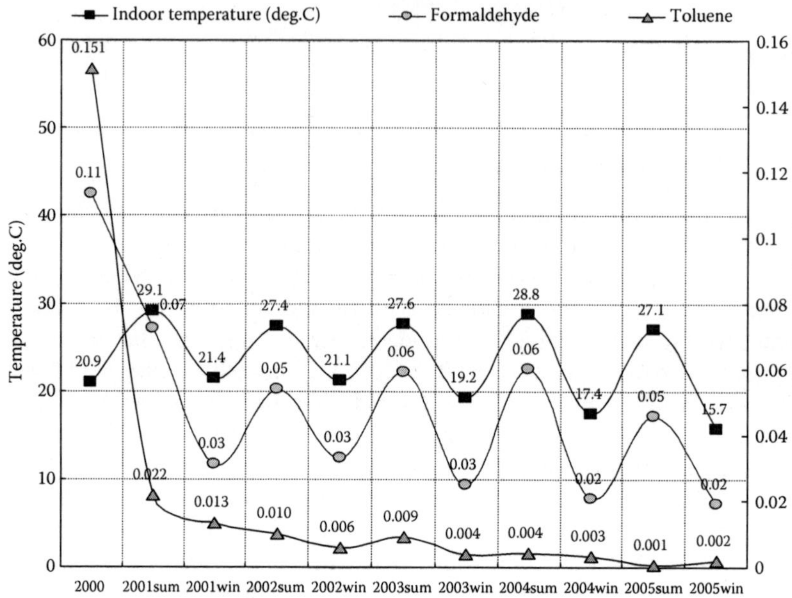

FIGURE 10.5 Change of concentrations of indoor pollutants with temperatures compared with guideline values.

10.3 TYPES AND FEATURES OF VENTILATION SYSTEMS

10.3.1 Types of Ventilation

Ventilation methods can be divided into two types: natural and mechanical ventilation. Natural ventilation utilizes external wind or a stack effect, while mechanical ventilation utilizes electric power. Mechanical ventilation systems are effective for extracting contaminants emitted locally or for the continuous ventilation for the whole building. Table 10.2 describes the three types of mechanical ventilation methods based on the function of the fans used.

TABLE 10.2
Types of Mechanical Ventilation

	Supply Air	Exhaust Air	Internal Pressure	Outline
Balanced ventilation	Mechanical fan	Mechanical fan	Controllable	Planned ventilation can be carried out by utilizing the balance of mechanical fans.
Supply-only ventilation	Mechanical fan	Naturally	Positive pressure	Contaminants cannot enter from next zones because of the positive pressure.
Exhaust-only ventilation	Naturally	Mechanical fan	Negative pressure	Generally, this type is used as local ventilation.

TABLE 10.3
Types of Hybrid Ventilation

	Natural Ventilation	Mechanical Ventilation	
Natural and mechanical ventilation	Independently	Independently	This principle is based on two fully autonomous systems in which the control strategy either switches between the two systems, or uses one system for some tasks and the other system for other tasks.
Fan-assisted natural ventilation	Main	Assist	This principle is based on a natural ventilation system combined with an extract or supply fan.
Stack and wind-assisted mechanical ventilation	Assist	Main	This principle is based on a mechanical ventilation system that makes optimal use of natural driving forces.

Source: Per Heiselberg: Principles of Hybrid Ventilation.

On the other hand, natural ventilation methods could also be effective for well-planned whole building ventilation. Natural ventilation, including a hybrid ventilation system that utilizes mechanical ventilation when the naturally driven sources cannot be used, is important from the viewpoint of energy conservation. Hybrid ventilation systems can be classified into several types based on the ventilation principles as shown in Table 10.3.

10.3.2 Methods of Selecting a Mechanical Ventilation System

Various kinds of ventilation fans are used for mechanical ventilation, so it is important to make a selection based on the site of installation or the purpose.

1. Common ventilation fans. These types of fans are representative and used in many places as exhaust fans. They are mostly propeller fans. Generally, they produce air flows with high volume and low pressure. Thus, in many cases they are combined with short ducts, and are installed mainly on walls that directly face outdoors.
2. Duct ventilation fans. These types mainly employ multiblade fans, and compared with propeller fans they create airflows with higher pressure. They are used particularly in medium- and high-rise buildings such as an apartment house. They are also broadly used in single-family detached houses.
3. Range hood fans. These types of fans are commonly used for local ventilation in kitchens and are installed just above ovens or stoves. They use multiblade fans, which provide airflows with high pressure.

4. Ventilation fans employed in Heating, Ventilating, and Air Conditioning (HVAC) systems. These are systems that consist of ventilation fan and heating or cooling facilities. They generally employ a heat exchange element that executes heat recovery between the exhaust air and supply air.

10.4 TRENDS IN TECHNOLOGY DEVELOPMENT AND ENERGY-SAVING PERFORMANCE

In recent years, because of global warming and rising environmental protection awareness, it became necessary to reduce the energy consumption in households (particularly for heating and cooling). Buildings have become more airtight, insulated, and soundproof, thus enabling indoor environments to be unaffected by outside conditions throughout the year. However, it has been said that owing to the reduction of air infiltration, problems of indoor air pollution arose, which involved emissions of VOCs and formaldehyde emitted from new building materials. This was the reason why revisions of the Building Standard Law of Japan enforced mandatory installation of ventilation equipment beginning July 2003.

Against this background, energy consumption increased for the operation of whole building ventilation systems. Even though the energy necessary to operate ventilation systems is not so large compared with that at other facilities, assuming that the ventilation system is operated 24 hours a day, 365 days a year, an energy conservation plan has become more important.

10.4.1 Mechanical Ventilation System Utilizing Low-Power Input Motors

In general, ventilation systems that use a brushless DC (EC) motor require less power input than ventilation systems that utilize AC motors, and therefore allow reductions in energy necessary for their operation. From the calculated results of Specific Fan Power (SFP) by using speciation data of two different residential heat exchange balanced ventilation system on specifications, by using a DC motor, approximately 40% to 50% of power input can be reduced. Nonetheless, in recent years, the overall efficiency of AC motor systems also has been improving and shows similar trends toward energy saving.

10.4.2 Demand-Controlled Ventilation

This type aims at energy conservation, in other words, the reduction of heating or cooling energy by decreasing supply or exhaust air as needed. This system employs various sensors (for example, carbon dioxide, carbon monoxide, sensors for other gases, as well as human presence sensors) to determine the demand for ventilation. The amount of energy saved varies depending on the indoor conditions.

10.4.3 Mechanical Ventilation Systems with a Heat Exchanger

The balanced ventilation system with a heat recovery system can be an energy-saving system, because it can reduce ventilation load, especially during heating or cooling. This type is used in housing, because it can reduce draught by preheating the supply air during heating season. However, when there is no need for air conditioning, nor heat exchange, or to intake a large amount of fresh air such as in a "night purge," it is recommended to select a system with a bypass duct or one-sided fan operation to reduce the energy needed for operation and the pressure loss on the heat-exchange element. This can also lead to a certain amount of energy saving.

10.4.4 Hybrid Ventilation System

There are many kinds of hybrid ventilation systems that utilize a stack effect or external wind pressure as driving forces for natural ventilation. Typically during winter, hybrid ventilation systems that use a stack effect can reduce the driving power for ventilation. According to theoretical calculation results, the stack effect can be used for ventilation 10% to 20% of the time during the winter period in Tokyo without the assistance of mechanical fans. Hybrid ventilation systems that use external wind pressure can also be used throughout the year. For example, this kind of hybrid ventilation system can work in apartment buildings without the assistance of mechanical ventilation 50% of the time. In addition, dumpers are required to avoid an excess ventilation rate during strong winds.

10.5 IMPORTANT POINTS FOR PLANNING AND OPERATION OF VENTILATION SYSTEMS

10.5.1 Planning of the Ventilation Path

When planning ventilation, it is necessary to consider and plan the layout of the ventilation path, because depending on the location of the pollutant there will be a need to change indoor air pressure to either positive or negative between zones.

For example, in areas such as a bathroom where contaminants (odor/moisture) might be generated, the air pressure must be kept negative to prevent the flow of the pollutant into other zones. On the other hand, in zones where there are no specific pollutant sources, the air pressure should be kept positive to prevent entry of polluted air from surrounding zones. However, when a supply-only ventilation system is employed, it is necessary to prevent the entry of moisture into external walls to prevent condensation. Notes for planning of the ventilation path are shown in the followings.

1. Whole building ventilation path

   It is necessary to consider the ventilation paths in the whole building by planning zones as clean zones (living room, bedroom, study, etc.) and dirty zones (kitchen, toilet, bathroom), and make the air flow from the clean zones into the dirty zones.
2. Planning the ventilation path in each rooms
   a. Living rooms, bedrooms, and similar habitable rooms
      - Extract terminals should be separate from supply terminals to decrease short-circuit problems.

- Supply terminals and air inlet preferably should be installed close to heating devices, in closets.
- In rooms with a large amount of pollutant emissions, local exhaust fans should be installed and exhaust the air directly into external.

b. Lavatories
- When a local exhaust fan system is employed, the airflow from other habitable rooms should be taken in through the lower parts of the door and the exhaust air directed into the exterior.

c. Kitchens

Odours, vapor, and combustion air are emitted intermittently in kitchens. Therefore a local exhaust ventilation fan is utilized to ensure a high ventilation volume in a short period of time. Nonetheless, considerations should be made so that the impacts for the ventilation of the whole building are kept to a bare minimum. In recent airtight housing, the indoor environment may be disturbed by excessive negative pressure caused by a lack of intended compensatory air. To avoid such cases, a simultaneous air supply and exhaust within the same zone is necessary. Moreover, there is a need for improvements in location and shapes of supply air terminal devices, as well as improvements that would prevent the disturbance of the indoor thermal environment.

d. Bathrooms and shower rooms

Because a large quantity of vapor/steam is emitted, it is beneficial to install exhaust ventilation fans with a dedicated duct.

10.5.2 Notes for Calculating Pressure Drops

When a duct type ventilation system is chosen, a ventilation fan should be selected by calculating pressure drops of the whole ductwork including terminal devices. In that case, if an excessively high safety factor were to be set, the airflow volume would also become excessively high. Thus, it is necessary to use measurement results based on reliable measurement methods in regards to pressure drop and airflow rate of ventilation system components.

10.5.3 Notes on Construction

When using flexible ducts, some bends, which are not intended but made during construction, are often found, caused by the good workability. Therefore, it is necessary to avoid higher pressure drops on the ducts caused by such issues when considering achieving the planed airflow rate on ventilation systems.

10.5.4 Notes Concerning Regular Maintenance

All components of ventilation equipment need to be installed in places from which airflow volume measurement and cleaning could be done easily. Cleaning is a vital part of regular maintenance because it can increase the energy efficiency of ventilation fans. The specific fan power and airflow rate were measured before and after

Step 1: Consideration of the ventilation system to choose

1) Confirm the lifestyle of the occupants, the planning of the house, the air tightness of the house, and the method used for heating and cooling.
2) Consider types of ventilation system and ventilation paths.

▼

Step 2: Planning for location of ventilation components

1) Consider the placement of outdoor and indoor terminal devices while taking into account the regular cleaning.
2) Consider the placement of the main unit of the ventilation system while taking into account the regular cleaning.
3) Consider the placement of the duct while taking into account the structure of the house.

▼

Step 3: Confirmation for ventilation system's performances taking account for energy conservation

1) Determine the design airflow rate.
2) Consider ways to minimize pressure loss at ducts and perform pressure loss calculations.
3) Select energy-efficient fans while considering their power consumption.

▼

Step 4: Confirmation for under/after construction

1) Confirm whether maintenance can be performed or not and make improvements.
2) Consider performing the airflow rate measurement.
3) Consider performing the airflow rate adjustment.

FIGURE 10.6 An example of steps for ventilation planning. (Courtesy of Building Research Institute, National Institute for Land and Infrastructure Management, Institute for Building Environment and Energy Conservation: Design Guideline for Low Energy Housing with Validated Effectiveness, 2010.)

cleaning for an exhaust-only ventilation system that had not been cleaned for 2 years. It was found that the specific fan power indicated for the power input required to carry the effective air was 1 m^3/h. The value before the cleaning was 1.3 times higher compared with the value after the cleaning. The airflow rate was 107 m^3/h and 137 m^3/h before and after cleaning, respectively. There is also a case in which the specific fan power in a balanced ventilation system rose approximately three times compared to that after 6 months of operation without cleaning.

10.5.5 Steps for Ventilation Planning

Figure 10.6 shows concrete steps for ventilation planning in housing. In addition to considerations for heating and cooling, the steps also consider the necessity for regular cleaning by occupants.

10.6 VENTILATION EQUIPMENT IN LARGE BUILDINGS

10.6.1 Ventilation in Designated Buildings

As described in Section 10.2, the Act on Maintenance of Sanitation in Buildings stipulates mandatory periodical measurement of air environments in large buildings, and thus ensures regular maintenance. These periodic measurements target the supply and exhaust fans of the ventilation systems employed in the centrally controlled

TABLE 10.4
Causes of Sick-Building Syndrome

- Inadequate ventilation
- Chemical contaminants from indoor sources
- Chemical contaminants from outdoor sources
- Biological contaminants

Source: EPA: Indoor Air Facts No. 4 (revised): Sick-Building Syndrome. Washington, DC: United States Environmental Protection Agency.

HVAC systems. Because levels of temperature, humidity, and ventilation operation were managed by the periodic measurements, Japan has managed to avoid serious SBS problems.

The United States Environmental Protection Agency shows the causes of sick-building syndrome (Table 10.4) and describes sick-building syndrome as follows: "The term 'sick-building syndrome' (SBS) is used to describe situations in which building occupants experience acute health and comfort effects that appear to be linked to time spent in a building, but no specific illness or cause can be identified. The complaints may be localized in a particular room or zone, or may be widespread throughout the building."

Various causes are listed in the table, but to summarize, sick-building syndrome can be attributed to contaminants resulting from the lack of planned ventilation. Through enforcement of laws and regulations such as the Act on Maintenance of Sanitation in Buildings, some countermeasures may be taken by utilizing results of periodic measurements of indoor air quality.

10.6.2 Ventilation Systems Except Those Employed in Centrally Controlled HVAC Systems

According to the Building Standard Law, ventilation systems except for those employed in centrally controlled HVAC systems include natural ventilation systems (as stipulated by Article 129:2, Section 6, Paragraph 1 of the Order) and mechanical ventilation systems (as stipulated by Article 129:2, Section 6, Paragraph 2 of the Order). However, mechanical ventilation is mandatory for habitable rooms and air purifying systems can be used. For these mechanical ventilations, it is necessary to calculate the effective airflow rate by using the following formula:

$$V = 20\,A_f \div N \tag{10.1}$$

where

V = Effective air flow rate
A_f = Floor area of target room
N = Substantial floor area for an occupant

The effective airflow rate for devices with air purifiers is calculated according to the concentration of formaldehyde before and after purification (as stipulated by Article 20:8 of the Order). Furthermore, if a single mechanical ventilation system supplies the ventilation for two or more habitable rooms, the required effective airflow rate needs to be higher than the sum of the effective airflow rate calculated separately for each room.

ENDNOTES

1. Osawa H, Hayashi M. 2009. Status of the indoor air chemical pollution in Japanese houses based on the nationwide field survey from 2000 to 2005. *Build Environ* 44:1330–1336.
2. Hayashi M, Osawa H. 2008. The influence of the concealed pollution sources upon the indoor air quality in houses. *Build Environ* 43:329–336.
3. Heiselberg P. 2002. *Principles of hybrid ventilation.* Aalborg, Denmark: Aalborg University.
4. Building Research Institute, National Institute for Land and Infrastructure Management, Institute for Building Environment and Energy Conservation. Design guideline for low energy housing with validated effectiveness, 2010.
5. United States Environmental Protection Agency. 1991. Indoor Air Facts No. 4 (revised). Sick Building Syndrome. Air and Radiation (6609J). Research and Development (MD-56).
6. Sawachi T, Osawa H et al. 2004. Airtightness of the envelope of wooden detached houses in Kanto area and the estimation of air leakage. *J Environ Enging* (Transactions of AIJ) 580:45–51. (In Japanese)

11 Ventilation, Air-Tightness, and Air Pollution

Hiroshi Yoshino, Dr. of Engineering, Professor Emeritus and President-Appointed Extraordinary Professor
Tohoku University

Rie Takaki, PhD, Associate Professor
Department of Life Design for Safety and Amenity, Faculty of Life Design, Tohoku Institute of Technology

Air-tightness of houses in Japan has become an efficient means of promoting reduction of air conditioning load and saving energy. It is easy to estimate that houses with high air-tightness without enough ventilation airflow rates produce more pollution of chemical substances in indoor air. Therefore we performed a measurement survey on chemical substance concentrations in indoor air, air-tightness, and air change rate to determine indoor air pollution due to chemical substances in actual houses, mostly those where sick-house syndrome was present [1].

11.1 OUTLINE OF THE MEASUREMENT SURVEY

11.1.1 Survey Period and Investigated Houses

We performed the measurement survey on chemical substance concentrations, air change rates, and air-tightness in 76 wooden detached houses of Tohoku Region from 2001 to 2008. Table 11.1 shows the survey period and the number of houses. The survey on air-tightness was performed in 54 houses and the one on air change rates in 39 houses with a mechanical ventilation system. In some houses surveys were performed more than once over several years. Figure 11.1 shows the types of ventilation systems. Forced supply and exhaust systems had been installed in 23 houses (30%), forced exhaust systems in 25 house (33%), and 28 houses (37%) adopted natural ventilation.

11.1.2 Survey Points and Measurement Method

The measurement was performed on ventilation airflow rates, air-tightness, and chemical substance concentrations. Table 11.2 shows the method of measuring chemical substances concentrations. Carbonyl compounds and volatile organic compounds (VOCs) were measured by sampling indoor air in two or three rooms of each house. Windows and doors of the rooms were closed as much as possible during

TABLE 11.1
Survey Period and Number of Houses

Measurement Year	Number of Houses	Number of Houses at Every Measurement Item: Air Tightness	Ventilation Rate (Airflow Rate at Exhaust of Forced Ventilation System)	Chemical Substance Concentrations: HCHO	TVOC
2001	27	27	12	25	25
2002	14	14	5	11	11
2003	9	9	7	9	9
2004	9	9	7	7	7
2005	19	9	13	9	9
2006	16	7	13	16	14
2007	5	5	0	5	5
2008	9	4	5	4	4
Total (First data)	108 (76)	84 (54)	62 (39)	86 (62)	84 (60)

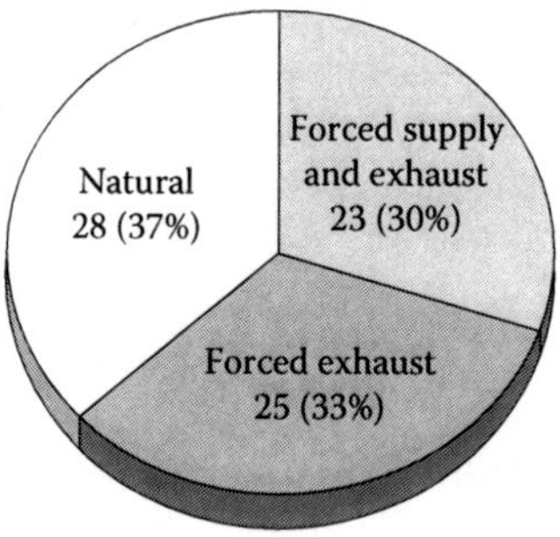

FIGURE 11.1 Types of ventilation systems.

TABLE 11.2
Measurement Method for Chemical Substance Concentrations

Items	Carbonyl Compounds (HCHO and Acetaldehyde)	VOCs (28 Substances)
Sampler	DNPH[a] cartridge (Sep-Pak XPoSure, Waters Co.)	Charcoal absorption (Jumbo Type, Sibata Scientific Tech. Ltd.)
Sampling method	Passive sampling (24 hours)[b]	Active sampling (300 mL/min, 24 hours)
Sampling point	1.2 m above the floor in 3 typical rooms (living room, bedroom, etc.)	
Analysis method	HPLC	GC–MS

[a] 2,4-Di-Nitro-Phenyl-Hydrazin.

[b] Active sampling was performed in 2008 (100 ml/min., 24 hours).

sampling for 24 hours. Temperature and relative humidity were measured using a small data logger with sensors (T&D, TR-72U). Air-tightness was measured by the depressurized method using an air-tightness measurement device (Kona Sapporo Co., KNS-400; Figure 11.2). The airflow meter (Kona Sapporo Co., Swema Flow 65; Figure 11.3) was used to measure the airflow rate at intake/exhaust of the investigated houses with the forced ventilation system installed. The airflow rate was

FIGURE 11.2 Air-tightness measurement device.

FIGURE 11.3 Airflow meter.

defined as the total amount of airflow rate at the exhausts divided by the total volume of the rooms.

11.2 MEASUREMENT RESULTS

11.2.1 Air-Tightness of Houses

Figure 11.4 shows the air-tightness of 54 investigated houses. The first measurement data are adopted for some houses with several measurements. The equivalent leakage area (hereinafter referred to as the leakage area) per unit floor area was distributed from 0.1 to 12 cm^2/m^2. According to the results, 43 houses (72%) satisfied the Energy Saving Standard, which requires the leakage area to be distributed less than

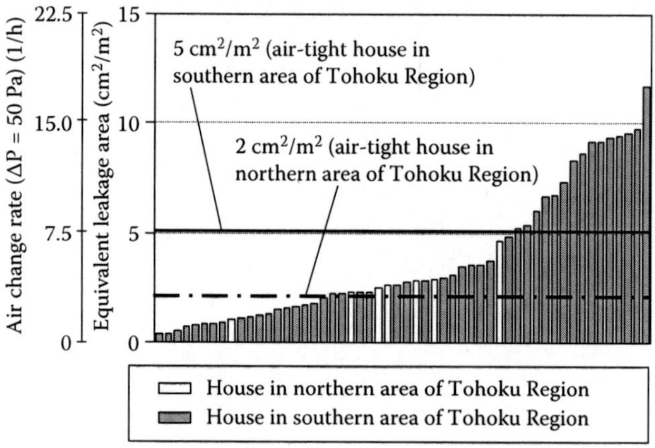

FIGURE 11.4 Air-tightness of houses (54 houses, first data).

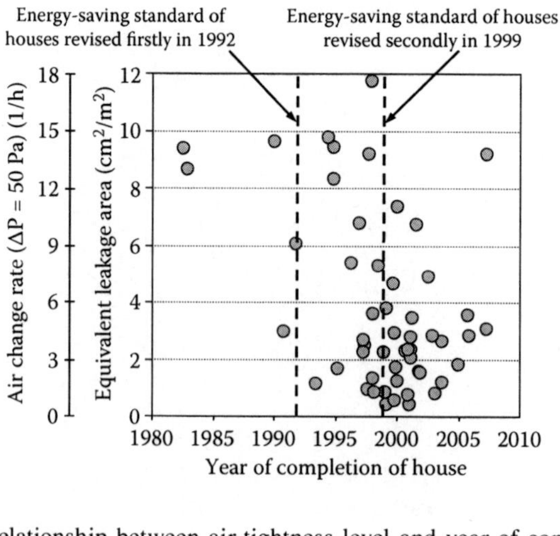

FIGURE 11.5 Relationship between air-tightness level and year of completion of houses (54 houses, first data).

2 cm²/m² in Hokkaido and three northern prefectures of Tohoku Region and less than 5 cm²/m² in other regions of Japan [2].

It is predicted that there are two major causes of differences in the level of air-tightness of the investigated houses: the years the houses were built and the age of the houses. Figure 11.5 shows the relationship between air-tightness level and the years the houses were completed. The investigated houses were built from 1982 to 2007. The air-tightness tends to be higher in newer houses, which reflects the progress of the air-tightness of Japanese houses. Figure 11.6 shows the relationship between air-tightness and the age of the houses. The age of the investigated houses distributes extensively over 20 years. The air-tightness tends to be lower in older houses.

The air-tightness measurements were performed several times over a few consecutive years in 14 houses. Figure 11.7 shows the results. The first measurement was performed 5 months to 18.5 years after the houses were built. The houses with high

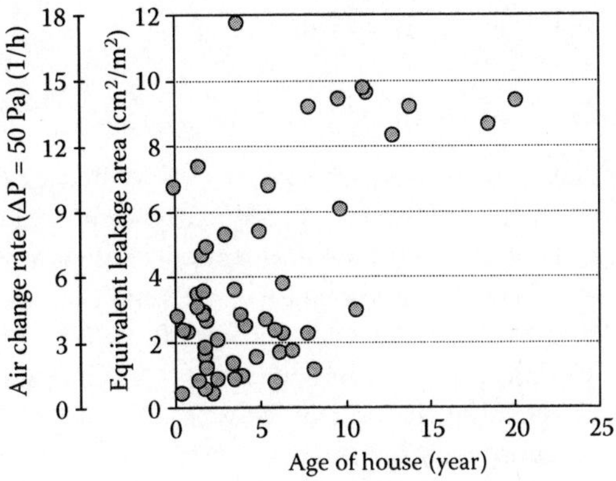

FIGURE 11.6 Relationship between air-tightness and age of houses (54 houses, first data).

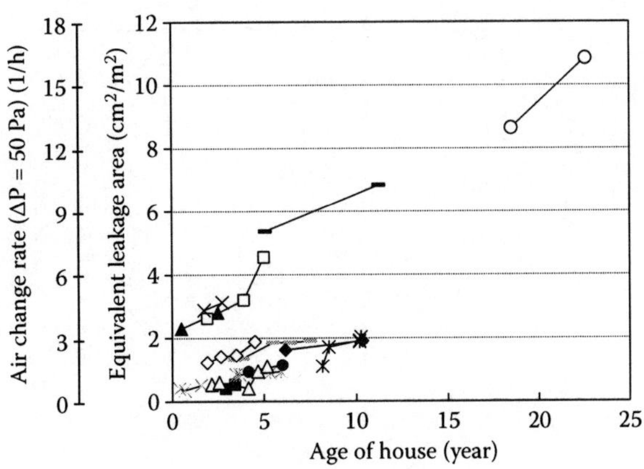

FIGURE 11.7 Long-term changes of air-tightness of houses (54 houses).

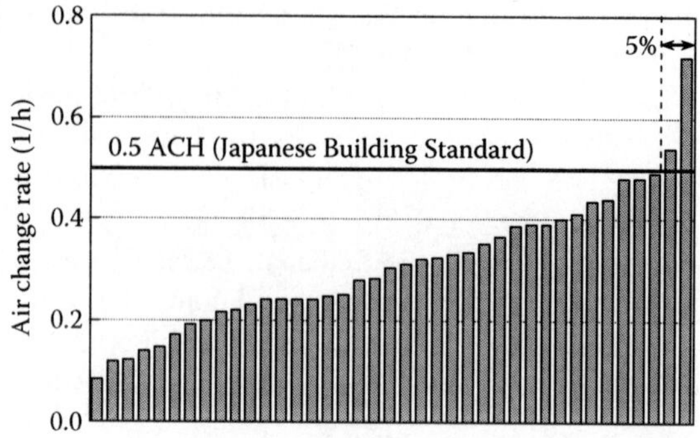

FIGURE 11.8 Air change rate of houses (39 houses, first data).

air-tightness level remain unchanged. Some other cases are reported in which the air-tightness doubled one year after completion [3,4] and the leakage area increased 100 cm^2/m^2 in one year [5].

11.2.2 Ventilation Rate of the Ventilation System

Figure 11.8 shows the results of air change rate of 39 houses (the first measurement data) measured by an airflow meter. Generally an air change rate of 0.5 air changes per hour (ACH) was used as the designated ventilation rate in many houses with a mechanical ventilation system installed, even built before 2003, the year of the Amended Building Standard Law. However, only two (5%) of the investigated houses could meet the designated ventilation rate of 0.5 ACH, and it is indicated that the ventilation rate is not sufficient in many houses.

Air change rate measurements were performed several times in nine houses. Figure 11.9 shows the results. In the house no. 5, the third measurement rate increased compared to the first and the second ones, and in house no. 7 the second one is higher than that of the first one. It is because a new ventilation system was installed before

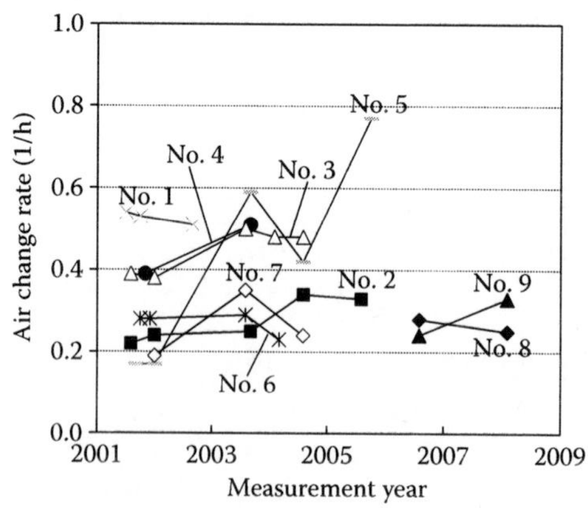

FIGURE 11.9 Long-term changes of air change rate of houses (9 houses).

the third measurement for house no. 5 and before the second measurement for house no. 7. The air change rate in house no. 5 increased to more than 0.5 ACH by installation of the new ventilation system. The air change rate of houses no. 2, 3, and 4 was increased after changing the driving mode from "low" to "high." The airflow rate decreased gradually in house no. 1 and at the fifth measurement of house no. 2; the second, fourth, and fifth of house no. 3; the fourth of house no. 5; and the third of house no. 7. It was found that one of the causes of insufficient ventilation rate was due to the problem related to the maintenance of the ventilation system such as choking air filters and insect screens covered by dust and insects. Furthermore, the results of this survey show that airflow rate decreases as the interval for cleaning a filter becomes long, and that air change rate becomes half when the ventilation filters are not cleaned for one year. The air change rate of house no. 5 increased greatly after cleaning the intake/exhaust of the system.

11.3 COMPARISON OF AIR-TIGHTNESS, VENTILATION SYSTEM, AND CHEMICAL SUBSTANCE CONCENTRATIONS

11.3.1 Comparison of Air-Tightness and Chemical Substance Concentrations

Figure 11.10 shows the relationship between air-tightness and formaldehyde concentration in 219 rooms of 77 houses, and Figure 11.11 shows the same relationship with total volatile organic compound (TVOC) concentration in 214 rooms of 77 houses. In the figures, the value of air-tightness of a house is shown on the *x*-axis and the chemical substance concentrations measured at the same term of each room in the same houses are plotted several points on the *y*-axis. Formaldehyde and TVOC concentration tend to be higher in high air-tightness houses. However, the house (*) in

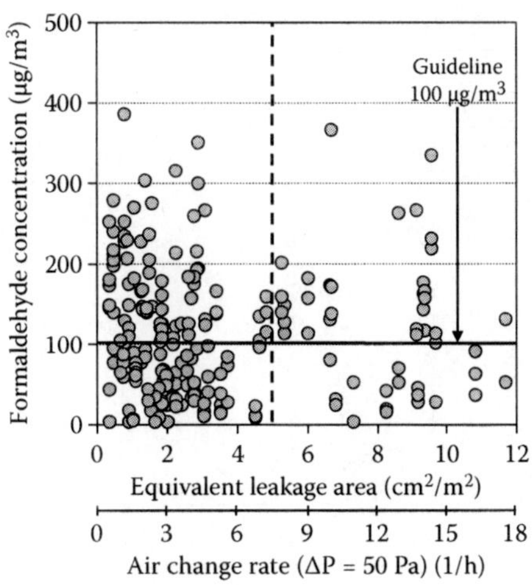

FIGURE 11.10 Relationship between air-tightness and formaldehyde concentration (219 rooms of 77 houses).

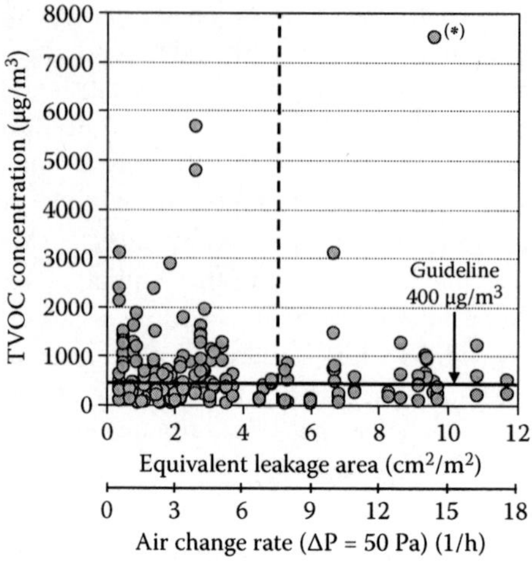

FIGURE 11.11 Relationship between air-tightness and TVOC concentration (214 rooms of 77 houses).

Figure 11.11 has high TVOC concentration in spite of low air-tightness and seems to be influenced by the emission rates of *p*-dichlorobenzene (7365 μg/m³) from mothballs. On the other hand, it is considered that the ventilation rate is also one of the influential factors.

11.3.2 Comparison of Air Change Rate and Chemical Substance Concentrations

Figure 11.12 shows the relationship between air change rate of 42 houses with a mechanical ventilation system and formaldehyde concentration in 110 rooms, and Figure 11.13

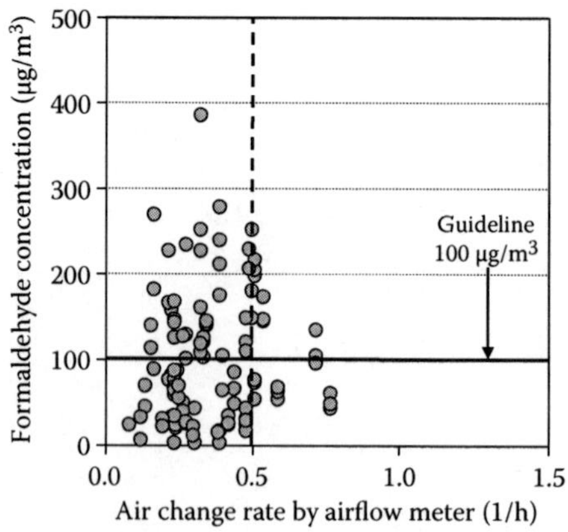

FIGURE 11.12 Relationship between air change rate and formaldehyde concentration (110 rooms of 42 houses).

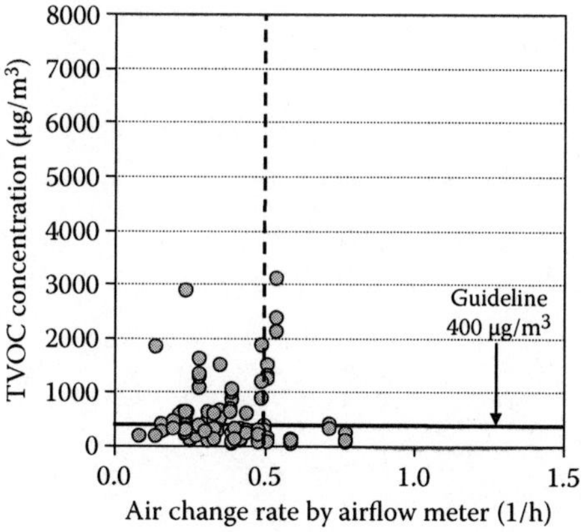

FIGURE 11.13 Relationship between air change rate and TVOC concentration (109 rooms of 40 houses).

shows the relationship between air change rate of 40 houses with a mechanical ventilation system and TVOC concentration in 109 rooms. The concentration tends to be higher in the rooms where air change rate is less than 0.5 ACH. However, the concentration can be higher than the guideline value in spite of a high air change rate, more than 0.5 ACH. We considered it to be influenced by pollutant sources.

11.4 CONCLUSIONS

The measurement survey revealed the fact that air-tightness of houses is improving but ventilation sufficiency is not secured. It has been confirmed that formaldehyde and TVOC concentration tends to be higher in high air-tightness houses and in rooms with a ventilation system of less than 0.5 ACH.

ENDNOTES

1. Yoshino H, Takaki R, Nakamura A, Ikeda K, Nozaki A, Kakuta K, Hojo S, Yoshino H, Amano K, Ishikawa S. Field Survey on Air-Tightness, Ventilation Rate and Indoor Air Quality of 77 Houses in Japan. In *Proceedings of 30th Air Infiltration and Ventilation Centre Conference*. Berlin, Germany. October 1–2, 2009.
2. The standard related air-tightness of house was deleted when the Energy Saving Standard was amended in 2009.
3. Elmroth A, Logdberg A. 1981. Airtight houses and energy consumption. *Building research and practice* 9(2):102–117.
4. O'Sullivan P, Jones PJ. 1982. The ventilation performance of houses: A case study. In *Proceedings of 3rd Air Infiltration Centre Conference*. London. September 20–23, 1982.
5. Irie Y, Fukushima A, Suzuki K, Ukita T, Konishi K. 1990. Air tightness of single detached houses in Hokkaido. In *Summaries of Technical Papers of Annual Meeting of Architectural Institute of Japan*. Hiroshima, Japan. October 1990. D:565–566 (In Japanese).

12 Chemical Features of Indoor Pollutants and Current Regulations

Naohide Shinohara, PhD, Senior Researcher
National Institute of Advanced Industrial Science and Technology

12.1 USAGE AND SOURCES OF CHEMICAL SPECIES DETECTED IN INDOOR ENVIRONMENTS

Many artificial chemical compounds have been used in various indoor products (Figure 12.1). Although these substances make our lives convenient, they could damage our health. Here, we discuss the typical chemical substances emitted from indoor materials.

Some examples of measurement data on the emission from rates indoor materials, such as furniture and building materials, to the indoor environment are shown in Figure 12.2. The results on the emission rates measured using a passive flux sampler (PFS) or passive emission colorimetric sampler (PECS) show that there are many emission sources of many kinds of chemical substances in the indoor environment. The measuring apparatuses, such as PFS and PECS, are explained in Chapter 13.

Measured indoor concentrations are compared with estimated indoor concentrations according to the emission rates and air exchange rates for typical chemical substances in a room (a) and formaldehyde in several houses (b).

12.1.1 Artificial Wooden Boards and Adhesives

Artificial wooden boards, such as plywood, particle boards, fiberboards, and laminate lumber, are one of the materials most used as building materials and furniture. As it is necessary to use old big trees when a large board for a wall or a door is made with plain wood, the cost is quite high and mass production is difficult. Therefore, an artificial wooden board was developed to obtain many large boards at low cost. Artificial wooden boards such as plywood began to be widely used during rapid economic growth in Japan (1960s). Plywood is manufactured by gluing thin wood plates peeled off from wood together with adhesive (resin) (Figure 12.3a). A particle board is manufactured by mixing wood particles or flakes together with adhesive and forming into a board (Figure 12.3b). Fiberboard is manufactured by mixing wood fibers, pulped (extracted) from wood chips with pressurized steam, with adhesive

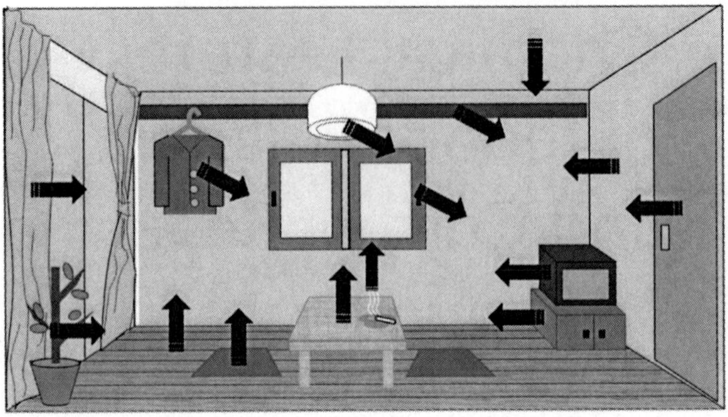

FIGURE 12.1 Various chemical compounds are emitted from various products in the indoor environment.

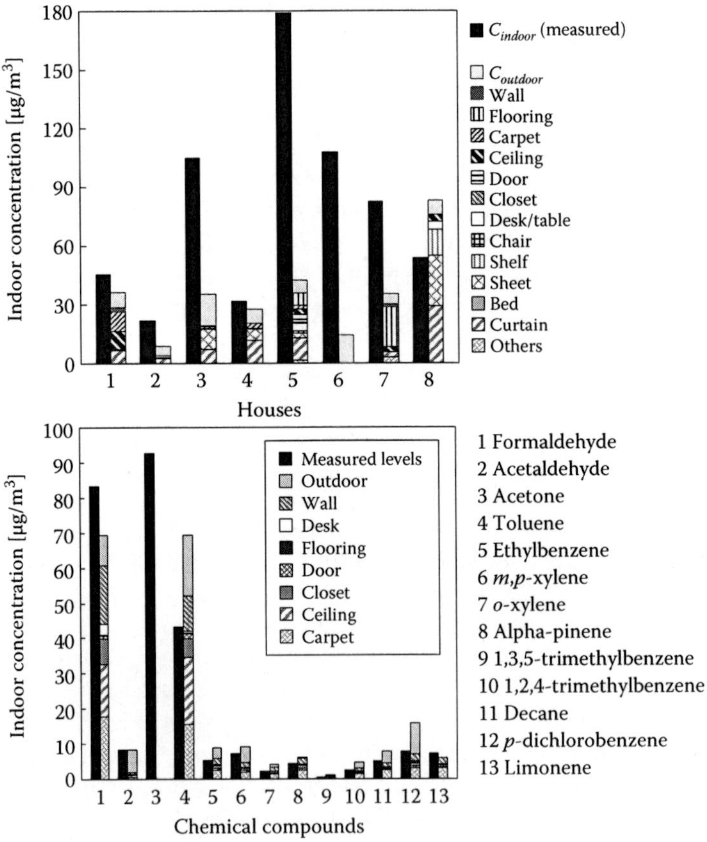

FIGURE 12.2 Contribution of each indoor source to the indoor concentrations.

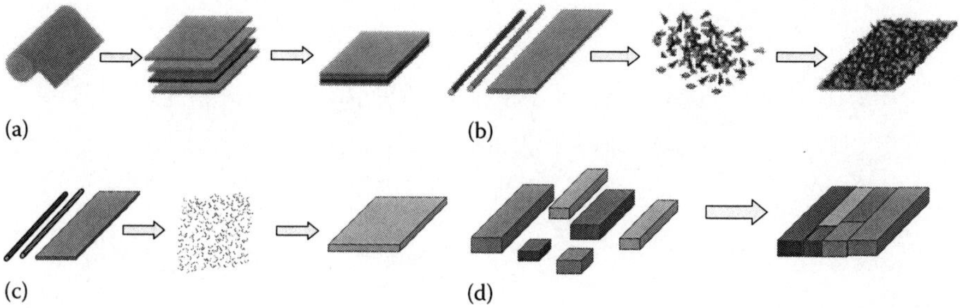

FIGURE 12.3 Conceptual diagram of manufacturing process of wooden board. (a) Plywood, (b) particle board, (c) fiber board, and (d) laminated lumber.

and forming into a board (Figure 12.3c). Laminate lumber is manufactured by bonding pieces of wood, such as scrap wood and end material, with adhesives (Figure 12.3d).

One of the primary emission sources of formaldehyde, which was the initial focus as the chemical responsible for the sick-house syndrome, identified in the late 1990s, was the adhesives in wooden materials. Because formaldehyde is the raw material of urea resin and urea melamine resin, and so forth, the unreacted formaldehyde may remain in these resins. Moreover, formaldehyde is generated by the hydrolysis of indoor moisture with urea resin (Figure 12.4). Residents were exposed to formaldehyde emitting from artificial wood in the indoor environment. As the residual amount of formaldehyde decreases over time, the emission rate can also decrease over time. However, formaldehyde is emitted even from old wooden board because the urea resin could continue to generate formaldehyde due to hydrolysis as long as resin and moisture exist. Because the emission rates of formaldehyde from the building materials were regulated by the Building Code revised in 2003, at present most wooden building materials rarely contain and emit formaldehyde at high levels in Japan. However, construction material is not covered by this regulation. Thus, high formaldehyde emission from the construction materials could pollute the indoor air through the outlet on the wall or gap of the walls [1]. Moreover, although the adhesives used for sticking wallpaper on the wall were also a typical source of the formaldehyde in the indoor environment, currently adhesives rarely contain and emit formaldehyde because of regulations such as the revised Building Code.

It is also reported that from a wooden board or raw wood itself, naturally produced chemicals such as terpenes (α-pinene, β-pinene, 3-caren, and a limonene),

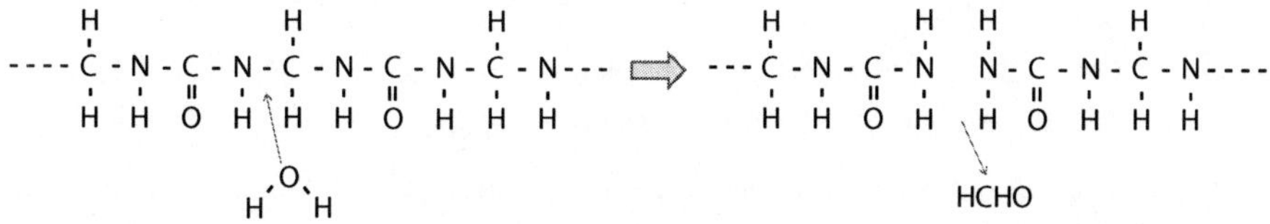

FIGURE 12.4 Hydrolysis of urea resin. Formaldehyde is generated by reacting urea resin of adhesive with indoor moisture. Because this reaction continues as long as the urea resin remains, formaldehyde can emit from old plywood.

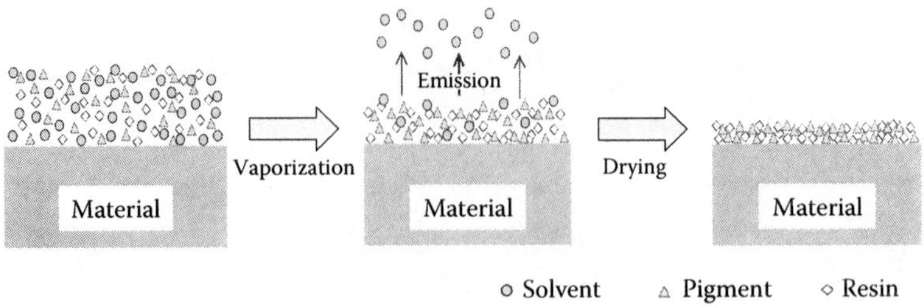

FIGURE 12.5 Principle of paint and vaporization of a solvent. After applying paint including resin, pigment, and solvent, only the solvent vaporize and emit. Pigment and resin coat are formed on the painting subject.

aldehydes (formaldehyde, acetaldehyde, etc.), and alcohols (ethanol, etc.) can be emitted. Therefore, acetaldehyde concentration in indoor environment exceeds the guideline value of the Ministry of Health, Labor and Welfare, Japan, in approximately 10% of new houses. Moreover, terpenes or alcohols could generate harmful secondary products in the indoor environment.

12.1.2 Paint

Chemical substances in paint solvent could emit from painted walls, ceilings, floor, and furniture. Paint consists of pigments that give color, resins that form a surface film, and volatile solvents that dissolve pigment and resin. After painting, because the volatile solvent evaporates, the pigment and resin form a fixed protective layer on the surface (Figure 12.5). Volatile organic solvents such as toluene, xylene, and ethylbenzene, which had been used as solvents, were emitted from paint until about 10 years ago. The mixing ratio of toluene and xylene was changed according to the indoor temperature or the humidity for uniform and smooth painting by controlling the vaporizing rate. After the guideline value set by the Ministry of Health, Labor and Welfare in 1997–2002, use of these organic solvents decreased sharply. Recently aqueous and organic solvents not listed in the guideline that contain esters such as ethyl acetate and butyl acetate and alcohols such as ethanol have been often used as solvents and emitted to the indoor environment from painted materials.

When a solvent volatilizes completely, the emission ends. This is different from the emission of formaldehyde from adhesives. Therefore indoor pollution due to painting solvents often disappears by 6 months to 1 year after the painting is completed.

12.1.3 Fungicides, Repellents, Mothballs, Air Fresheners, and Disinfectants

Although traditional Japanese houses have eaves that open wide and provide good air permeability, the structure of the foundation was mandated in residential houses after the end of World War II as a countermeasure against earthquakes. As high temperature and high humidity can easily permeate the closed areas under the eaves in Japan, preservatives, fungicides, and repellents have been used in building materials, especially in wooden structural materials to prevent wood rot and termite damage. Organochlorine compounds, such as chlordane, and organophosphorus compounds, such as chlorpyrifos, that had been widely used as termite control chemicals at one time were forbidden in 1986 and 2003, respectively. Currently, the most general termite control method is chemical infusion into wood under increased pressure, such as quaternary ammonium compound, copper ammonium compound, copper azole compound, and metal salt of fatty acid. As these chemicals are less volatile, concentrations indoors and under eaves are quite low [2]. Pyrethroids, which have a strong repellent effect against termites, and phenols applied or sprayed to wood could pollute the indoor environment.

In a high-temperature and high-humidity indoor environment, condensation could occur on the bottom of tatami and mold may grow. Thus, according to the Japanese Industrial Standards (JIS) [3], a suitable insect control treatment should be installed at the bottom of the tatami to prevent invasion by mites and other insects. An insect control sheet containing volatile chemicals, such as pyrethroid, organophosphorus compounds, and naphthalene, may be used. In such cases, these substances are emitted to the indoor air, which may become polluted.

Alcohols, aldehydes, imidazoles, and so on are used as antifungal agents in building materials. Although until 1990s formaldehyde contained in the adhesives of plywood or wallpaper prevented mold, an increase of mold in indoor environments has been reported after the regulation for formaldehyde was included in the Building Code.

In a high-temperature and high-humidity area such as Japan, moth repellent is indispensable to save clothes from insects. Tablet type and hanger type moth repellents are used in cloth storage cases and drawers. The moth repellent camphor, *p*-dichlorobenzenes, and naphthalenes, which evaporate to the air in the storage cases and are adsorbed on clothes, had been used often. Recently, use of pyrethroid as a moth repellent is increasing because this chemical has no odor.

12.1.4 Plasticizers and Flame Retardants

Plasticizers are additives that provide flexibility to plastics by disturbing the regular structure of a resin (Figure 12.6). Plasticizers are added to many kinds of plastics, such as polyvinyl chloride (PVC), and are detected in the indoor environment. Phthalates, such as di-isononyl phthalate (DiNP) and diethylhexyl phthalate (DEHP), and phosphates, such as tricresyl phosphate (TCP), are widely used as plasticizers for plastics. These chemicals have high vapor pressure and are easy to adsorb on the

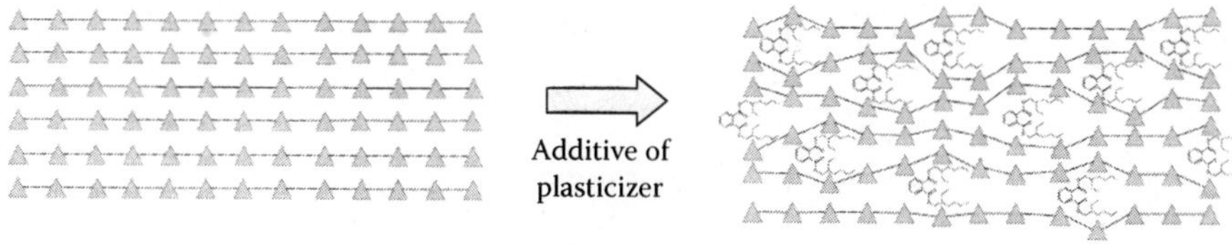

FIGURE 12.6 Principle of plasticizers. Because plasticizer penetrates into the resin structure and disturbs the structure, the resin is plasticized.

surface of flooring and dust. These chemicals are present in a much higher abundance in house dust than in indoor air.

Flame retardants are additives that provide fire resistance to consumer products such as plastics, rubber, fiber, paper, and wood. Although inorganic oxides are most commonly used as flame retardants, residences are rarely exposed to them because they do not evaporate. Bromine compounds, such as decabromodiphenylether, and phosphates, such as triphenyl phosphate, which inhibit burning by capturing free radicals and delay oxygen supply, could evaporate and pollute the indoor air.

12.1.5 Smoking and Burning Appliances

The side stream smoke from tobaccos is also a large source of indoor air pollution. The side stream of cigarettes contains many kinds of pollutants, such as particulate matter, carbon monoxide, nicotine, acetaldehyde, formaldehyde, benzene, nickel, and cadmium at high concentrations [4]. The concentrations of a large number of chemicals are higher in the side stream than those in the main stream. In the smoking room, there is a strong correlation between the concentration of nicotine and those of the volatile organic compounds (VOCs), such as formaldehyde, acetaldehyde, and benzene [5].

Similarly, open type burning appliances, which can emit particulate matter, carbon monoxide, acetaldehyde, formaldehyde, benzene, and so on, also cause indoor air pollution. In particular, in developing countries, many residences use biomass fuel or coal in indoor environments. Open burning of biomass fuel and coal generates these chemicals at much higher levels than burning appliances, causing serious indoor air pollution. Moreover, the indoor nitrogen dioxide and carbon monoxide in the temporary houses after a great earthquake in east Japan at 2011 are very high concentrations, because of the small sizes of the rooms and low ventilation rates [6].

12.1.6 Secondary Generated Compounds

The concentration of formaldehyde and acetaldehyde rises by reactions of styrene and 4-vinyl cyclohexane emitted from carpets with ozone [7]. The concentration of aldehydes and particulate matter also rises as a result of the reaction of α-pinene and limonene emitted from raw wood with ozone [8]. Most indoor ozone comes from outdoor air, and the indoor ozone concentration is 20% to 70% of the outdoor level [9]. In the indoor environment, some laser printers and electrostatic air cleaners can generate ozone [10].

Moreover, it has been widely known that 2-ethyl 1-hexanol is generated by the reaction of DEHP, which is the most typical phthalate with indoor water. Especially in reinforced concrete buildings, moisture in concrete reacts with DEHP in PVC flooring on the concrete floor and 2-ethyl 1-hexanol is emitted to indoor air [11].

12.1.7 Contaminants in Outdoor Air

In roadside houses and neighborhoods where manufacturing or incinerator plants are located, outdoor contaminants could enter the indoor environment through ventilation.

Although outdoor pollution has decreased in Japan in recent years, outdoor air is seriously polluted in some developing countries.

Some chemicals enter the indoor environment through adsorption on clothes. In farmers' houses, the indoor organophosphorus pesticide concentration is reported to be increased after spraying of pesticides in outdoor farms. This could be because of the adsorption of pesticide on clothes. Similarly, the concentration of organochlorine compounds increases after picking up clothes from the dry cleaners.

12.1.8 Unknown Pollutants

Recently, indoor concentrations of typical aldehydes and VOCs have been decreased by setting the guideline value for indoor air pollutants and revision of the Building Code. On the other hand, indoor levels of alternative chemicals could be increasing, although some of the alternative chemicals cannot be identified. Some chemicals might induces adverse effects at low levels in the indoor environment.

12.2 DETERMINATION OF POLLUTANTS FROM THE VIEW OF HEALTH EFFECTS

The adverse health effects caused by the indoor air pollutants are explained as follows.

12.2.1 Target Health Effects

12.2.1.1 Sick-House Syndrome

After the 1970s, office workers complained of symptoms such as mucosal irritation, skin irritation, fatigue, headache, and impaired concentration in newly built air-tight office buildings, and the condition was called sick-building syndrome (SBS) [12]. In Japan, since the building administration law required the carbon dioxide levels to be 1000 ppm or less in buildings with 3000 m^2 or more of floor area, high ventilation had been maintained and SBS did not appear in office buildings. After the second half of the 1990s, however, air-tightness of residential houses was improved to save energy consumption, many residents complained of symptoms similar to SBS in new residential houses, and it became a social problem called sick-house syndrome (SHS). Moreover, in classrooms of newly built schools, some students or teachers also complained of these symptoms

The symptoms of SHS vary among different people, although characteristic symptoms are mucosal irritation, skin irritation, fatigue, headache, and impaired concentration. Formaldehyde used in the adhesives and toluene and xylene used as solvents were the main causative agents of SHS. The indoor concentrations of these chemical compounds sharply decreased as a result of guideline values for 13 kinds of chemicals set by the Ministry of Health, Labor and Welfare and revision of Building Code by the Ministry of Land, Infrastructure, Transport and Tourism. Accordingly, the SHS problem has receded recently. However, SHS caused by new alternative chemicals such as 1-methyl 2-pyrolidone and texanol in water paint has been reported [13].

12.2.1.2 Asthma

Asthma is a common allergic inflammatory disease in the airways caused by the deposition of chemical compounds and particles. Wheezing, coughing, chest tightness, and shortness of breath are the major symptoms of asthma. In an indoor environment, nitrogen dioxide, some chemical compounds, mold, and mites are considered to be the responsible substances. Recently, an increasing number of reports have indicated that the indoor concentrations of phthalate esters could correlate with asthma incidence. In developing countries, asthma is often induced in the indoor environment by the combustion of biomass fuel and coal.

12.2.1.3 Endocrine Disruption

Some indoor chemicals could disrupt the development of the endocrine system by their hormone-like action. Endocrine disruption became a popular topic after publication in 1996 of "*Our Stolen Future*" by Theo Colborn in 1997 [14]. Since phthalate esters, such as diethylhexyl phthalate (DEHP), dibutyl phthalate (DBP), diisobutyl phthalate (DiBP), and the benzyl butyl phthalate (BBP), which were often detected in the indoor environment, could be the endocrine disrupting chemicals, use of these chemicals was limited in some countries such as Denmark. On the other hand, many reports claim that phthalate esters, such as DEHP, have little or no endocrine disrupting capacity.

12.2.1.4 Cancer

In Japan, cancer is responsible for more than 30% of all deaths [15]. Some of the cancer cases could be caused by exposure to chemical compounds. In the indoor environment, formaldehyde and benzene are carcinogenic and acetaldehyde, ethylbenzene, and *p*-dichlorobenzene are possibly carcinogenic. Formaldehyde is emitted from the adhesives described earlier. Although most indoor benzene emitted from burning instruments such as heating instruments, or introduced from tobacco smoke or outdoor air in Japan (almost no indoor emission source), benzene is also used as a painting solvent in some countries such as China. Moreover, it is supposed that polycyclic aromatic (PAHs) generated from burning instruments are probably also carcinogenic.

12.2.2 Field Survey of Indoor Pollution

The Ministry of Health and Welfare (present, Ministry of Health, Labor and Welfare) conducted a survey to measure the indoor concentrations of VOCs in residential houses throughout Japan in 1997 and 1998 (1997: 180 houses; 1998: 205 houses). The average concentration of toluene and *p*-dichlorobenzene was comparatively high, 93.3 and 128.4 μg/m^3 in 1997 and 98.3 and 123.3 μg/m^3 in 1998, respectively (Table 12.1) [16].

TABLE 12.1
Indoor Concentrations of Volatile Chemicals in Japan

	MHLW Survey (1999) [16]				NEDO Survey (2007) [17]									
	1997		1998		Summer 2005		Autumn 2005		Winter 2006		Spring 2006		Summer 2006	
Substances	Mean [μg/m³]	Median [μg/m³]	Mean [μg/m3]	Median [μg/m³]	Mean [μg/m³]	Median [μg/m³]	Mean [μg/m³]	Median [μg/m³]	Mean [μg/m³]	Median [μg/m³]	Mean [μg/m³]	Median [μg/m³]	Mean [μg/m³]	Median [μg/m³]
Formaldehyde	–	–	–	–	94.4	71.5	49.4	36.4	27.8	23.3	58.2	41.9	56.7	37.0
Acetaldehyde	–	–	–	–	47.5	31.1	37.9	29.2	43.1	29.8	38.6	27.9	29.2	12.1
n-Hexane	7.4	3.6	7.0	2.9	2.09	N.D. (0.21)	1.94	N.D. (<0.21)	10.8	4.66	9.46	2.56	6.69	1.85
n-Heptane	7.7	2.0	7.8	2.5	5.30	2.33	5.67	2.81	6.73	2.01	5.40	2.46	5.08	2.13
n-Octane	11.5	1.6	12.7	1.8	2.75	1.08	4.64	1.30	5.98	t.r. (<0.85)	3.25	t.r. (<0.85)	1.79	t.r. (<0.85)
n-Nonane	20.9	3.3	20.8	4.8	8.52	2.71	12.8	3.44	13.0	1.70	8.38	2.27	4.77	1.45
n-Decane	23.1	4.2	21	7.4	24.3	22.2	12.4	8.59	4.53	N.D. (<0.22)	6.79	3.49	5.51	2.90
n-Undecane	14.6	2.9	13	4.6	11.9	4.19	8.13	3.52	16.0	6.18	7.36	1.13	4.79	2.47
n-Dodecane	9.5	2.6	10.2	4.8	13.5	5.32	7.27	N.D. (<0.42)	2.95	N.D. (<0.42)	7.68	N.D. (<0.42)	5.99	3.48
n-Tridecane	7.3	1.9	13.1	5.7	–	–	–	–	–	–	–	–	–	–
n-Tetradecane	5.7	2.8	18.7	4.4	842	240	457	8.37	31.6	8.37	929	8.37	468	117
n-Pentadecane	2.0	0.5	5.3	1.4	–	–	–	–	–	–	–	–	–	–
n-Hexadecane	1.3	0.3	2.3	0.8	–	–	–	–	–	–	–	–	–	–
Benzene	5.9	3.1	7.2	2.6	1.71	1.58	2.90	2.46	3.60	2.92	2.01	1.84	1.82	1.51
Toluene	93.3	26.9	98.3	25.4	36.1	22.1	36.6	23.9	36.4	26.9	26.8	16.4	22.7	15.6

(Continued)

TABLE 12.1 (CONTINUED)
Indoor Concentrations of Volatile Chemicals in Japan

	MHLW Survey (1999) [16]				NEDO Survey (2007) [17]									
	1997		1998		Summer 2005		Autumn 2005		Winter 2006		Spring 2006		Summer 2006	
Substances	Mean [µg/m³]	Median [µg/m³]	Mean [µg/m3]	Median [µg/m³]	Mean [µg/m³]	Median [µg/m³]	Mean [µg/m³]	Median [µg/m³]	Mean [µg/m³]	Median [µg/m³]	Mean [µg/m³]	Median [µg/m³]	Mean [µg/m³]	Median [µg/m³]
Ethyl Benzene	21.6	6.8	22.5	6.8	11.9	7.41	11.0	8.75	7.42	5.67	19.2	7.24	6.74	5.60
m,p-Xylene,*m*,Xylene	26.7	9.6	24.3	10.2	24.8	15.7	24.7	19.1	22.9	12.8	17.7	13.3	13.4	10.5
o-Xylene	11.5	4.2	10	3.8	5.61	3.60	5.52	4.01	5.19	2.42	4.17	2.81	3.05	2.31
Styrene	4.5	0.6	4.9	0.2	t.r. (<1.1)	N.D. (<0.33)	t.r. (<1.1)	N.D. (<0.33)	N.D. (<0.33)	N.D. (<0.33)	2.67	N.D. (<0.33)	2.34	t.r. (<1.1)
1,3,5-Trimethylbenzene	9.6	1.3	3.1	1.3	1.75	1.13	1.66	0.515	1.09	N.D. (<0.21)	1.16	t.r. (<0.70)	0.983	t.r. (<0.70)
1,2,4-Trimethylbenzene	29	4.1	12.8	4.8	6.85	4.60	6.69	3.99	3.63	1.53	4.06	2.59	3.86	2.98
1,2,3-Trimethylbenzene	5.8	1.0	4.2	1.2	1.67	1.07	1.60	t.r. (<1.0)	1.08	N.D. (<0.34)	1.03	t.r. (<1.0)	t.r. (<1.0)	t.r. (<1.0)
alpha-Pinene	12.9	3.6	77.6	4.7	139	13.5	60.0	12.1	37.8	8.20	67.9	13.1	61.4	12.0
d-Limonene	26.5	6.1	42.1	12.8	32.9	17.6	25.3	16.2	12.7	7.40	49.6	12.5	32.1	14.3
Dichloromethane	7.5	2.6	–	–	206	1.15	161	2.39	95.9	1.81	44.1	1.50	28.5	0.954
Trichloroethylene	7.9	0.2	2.4	0.3	1.67	t.r. (<0.94)	3.42	2.26	2.89	2.12	1.45	1.11	1.71	t.r. (<0.94)
Tetrachloroethylene	1.8	0.4	1.9	0.3	1.01	t.r. (<1.0)	1.11	t.r. (<1.0)	t.r. (<1.0)	t.r. (<1.0)	1.15	t.r. (<1.0)	t.r. (<1.0)	N.D. (<0.32)
Chloroform	2.1	0.4	1.0	0.3	0.831	N.D. (0.19)	N.D. (<0.19)	N.D. (<0.19)	t.r. (<0.63)	N.D. (<0.19)	1.60	1.00	0.687	t.r. (<0.63)

(Continued)

TABLE 12.1 (CONTINUED)
Indoor Concentrations of Volatile Chemicals in Japan

	MHLW Survey (1999) [16]				NEDO Survey (2007) [17]									
	1997		1998		Summer 2005		Autumn 2005		Winter 2006		Spring 2006		Summer 2006	
Substances	Mean [μg/m³]	Median [μg/m³]	Mean [μg/m3]	Median [μg/m³]	Mean [μg/m³]	Median [μg/m³]	Mean [μg/m³]	Median [μg/m³]	Mean [μg/m³]	Median [μg/m³]	Mean [μg/m³]	Median [μg/m³]	Mean [μg/m³]	Median [μg/m³]
1,2-Dichloropropane	1.1	0.2	0.5	0.2	N.D. (<0.29)	N.D. (<0.29)	N.D. (<0.29)	N.D. (<0.29)	N.D. (<0.29)	N.D. (<0.29)	t.r. (<0.96)	N.D. (<0.29)	N.D. (<0.29)	N.D. (<0.29)
p-Dichlorobenzene	128.4	12.3	123.3	16.1	352	16.3	96.7	8.63	29.1	2.81	327.9	7.64	209.8	13.4
Tetrachloromethane	3.6	0.4	1.5	0.6	1.28	1.34	1.26	1.26	t.r. (<1.1)	t.r. (<1.1)	t.r. (<1.1)	1.10	t.r. (<1.1)	t.r. (<1.1)
Chloro Dibromomethane	5.3	0.2	2.0	0.2	t.r. (<1.1)	N.D. (<0.34)	t.r. (<1.1)	t.r. (<1.0)	t.r. (<1.1)	N.D. (<0.34)	t.r. (<1.1)	N.D. (<0.34)	t.r. (<1.1)	N.D. (<0.34)
Ethyl acetate	9.0	3.8	11.9	3.7	29.2	6.41	24.5	17.8	15.8	9.48	19.3	t.r. (<1.0)	16.6	8.74
Butyl acetate	10.3	2.1	11.7	1.9	22.3	2.84	5.09	1.85	2.14	1.30	5.00	2.77	5.76	2.23
Acetone	32.3	18.3	–	–	42.1	24.4	31.9	24.6	23.3	20.7	33.7	28.3	38.5	31.1
Methyl Ethyl Ketone	6.6	2.3	5.8	1.6	13.9	7.13	17.1	13.1	9.76	6.94	9.51	7.75	8.63	6.47
Methyl iso-Butyl Ketone	7.3	0.8	4.8	0.8	7.57	2.07	4.43	1.58	1.75	t.r. (<1.1)	4.00	1.61	2.97	1.25
Ethanol	281.2	84.7	–	–	58.0	20.9	326	180	198	140	131	60.8	90.6	43.9
i-Butanol	–	–	–	–	1.27	N.D. (<0.14)	0.726	N.D. (<0.14)	t.r. (<0.47)	N.D. (<0.14)	1.11	0.930	0.981	0.670
n-Butanol	5.1	1.9	6.8	1.4	4.03	2.37	1.60	1.19	0.798	0.617	2.46	1.66	2.76	2.11
Nonanal	5.9	3.4	15.8	6.8	52.7	49.0	23.8	20.0	24.8	20.7	70.1	64.8	50.5	46.8

In the survey conducted by the New Energy and Industrial Technology Development Organization (NEDO) from 2005 to 2006, the indoor concentrations of most chemicals were higher in summer and lower in winter (Table 12.1) [17]. The indoor concentrations of formaldehyde and toluene, which have decreased in new houses because of the guideline and revised Building Code, remained high in some houses built before the guideline was set.

12.3 RISK ASSESSMENT

12.3.1 What Is Risk Assessment?

"Risk of chemical compounds" represents the possibility and severity of adverse effects on human health or the environment due to the chemical compounds. According to the results of risk assessment, regulation and countermeasures will be decided or selected. A risk is evaluated by the combination of hazard and exposure. We can use highly toxic compounds if the possibility and amount of exposure are quite low and the benefit is quite high. On the contrary, even if the toxicity of the chemical compound is not so high, we do not have to use the chemical when the exposure levels are quite high. No chemical compounds have zero risk; we permit a certain amount of risk compared to the benefit. Therefore, the acceptable risk depends on the magnitude of the benefit. In the case of medication, the drug's efficacy is the benefit and the side effects are the risk. For cold medicines, because the benefit is not so large, only a minor side effect such as sleepiness is acceptable. On the other hand, highly adverse side effects can be accepted for anticancer drugs because the benefit is large.

12.3.2 Hazard Assessment

Hazard assessment is conducted based on epidemiological studies on humans or animal tests. Most chemicals do not show toxicity below a certain exposure level, which is called the threshold (for some diseases such as certain cancers, there is no threshold). Because it is difficult to obtain the threshold itself from the epidemiological study or animal test, the no observed adverse effect level (NOAEL), which is the highest exposure level without adverse effects, is used as the toxicological indicator. NOAEL is considered to be the same or lower than the threshold. In consideration of toxicity difference between humans and animals and among individuals and the reliability of the tests, the standard value or guideline value is obtained by dividing NOAEL by the uncertainty factor associated with these differences and reliabilities, as shown in Figure 12.7. As there are often several results of several toxicity tests even for a certain chemical, the lowest NOAEL is among the reliable tests for hazard assessment.

The indoor guideline values set by Ministry of Health, Labor and Welfare are decided based on the hazard assessment. As an example of hazard assessment, the procedures of hazard assessment for formaldehyde and *p*-dichlorobenzene are shown in Figure 12.7.

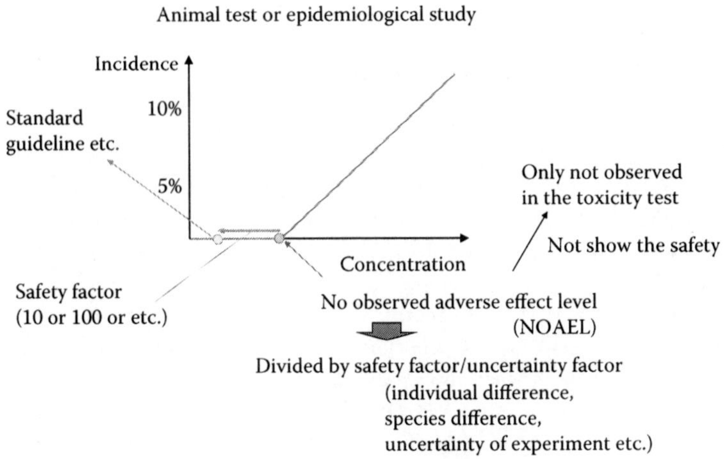

FIGURE 12.7 Hazard assessment of chemical compounds.

The indoor guideline value for formaldehyde is calculated based on the result of the nasopharyngeal mucosal irritation test in humans. It can be estimated that subjects did not feel irritation by exposure to up to 100 μg/m^3 of formaldehyde for 30 minutes, and is NOAEL for humans. In this case, because the test was conducted in humans and the adverse effects were not severe, no uncertainty factor was used and the guideline value for formaldehyde for 30-minute exposure was set at 100 μg/m^3.

In the case of *p*-dichlorobenzene, because there are no toxicity data on humans, the NOAEL is obtained from animal tests. In the repeat-dose 1-year study (5 days/week) on beagle dogs, an increase in liver weight and hypertrophy of hepatocytes were observed at levels higher than 50 mg/kg/day while no adverse effects were observed at 10 mg/kg/day. Therefore, the NOAEL of *p*-dichlorobenzene was estimated as 10 mg/kg/day on beagles. A correction for 7 days of exposure and differences of body weight and respiratory volume was performed for the NOAEL. The corrected NOAEL was divided by the uncertainty factor for the species difference, 10, and the individual difference, 10, and then the guideline value was decided to be 240 μg/m^3.

Guideline value = 240 μg/m^3
(= 10 × 7/5 × 50/15/100)
NOAEL
Correction of exposure duaration
(5 days → 7 days)
Correction of body weight
(per kg → 50 kg)
Correction of respiratory volume
(Human: 15m^3/day)
Uncertainty factor
(Individual difference 10)
(species difference 10)

TABLE 12.2
Indoor Guideline Values in Japan

Date	Chemicals	Guideline
June 1997	Formaldehyde	100 μg/m³
June 2000	Toluene	260 μg/m³
June 2000	Xylene	870 μg/m³
June 2000	*p*-Dichlorobenzen	240 μg/m³
Dec. 2000	Ethylbenzene	3800 μg/m³
Dec. 2000	Styrene	220 μg/m³
Dec. 2000	Chlorpyrifos	1 μg/m³ Child: 0.1 μg/m³
Dec. 2000	Di-*n*-butyl phthalate	220 μg/m³
Dec. 2000	Total VOC (TVOC)	400 μg/m³ [provisional]
July 2001	Tetradecan	330 μg/m³
July 2001	Diethylhexyl phthalate	120 μg/m³
July 2001	Diazinon	0.29 μg/m³
Jan. 2002	Acetaldehyde	48 μg/m³
Jan. 2002	Fenobucarb	33 μg/m³

For the other volatile chemicals, the guideline values were set by the toxicity assessment as well as formaldehyde and *p*-dichlorobenzene levels. Only for total VOCs, as we could not obtain NOAEL, the interim target value, not guideline value, was set to be 400 μg/m³, which was considered to be feasible from the national survey. The Japanese indoor guideline values are shown in Table 12.2 [18].

12.3.3 Exposure Assessment

Exposure assessment is the process of estimating/measuring how much/at what level and how many people are exposed to certain chemical compounds. Humans can be exposed to chemical compounds through three routes: (1) inhalation exposure, in which airborne chemical compounds are taken up through the respiratory tract and lungs from the mouth and nose during respiration; (2) oral exposure, in which food, drink, and dust are taken up through the digestive tract from the mouth by drinking and eating; and (3) dermal exposure, in which airborne, waterborne, and chemical compounds in products are taken up through the skin (Figure 12.8).

At first, the target group of exposure assessment, such as all Japanese or neighborhoods that have a certain kind of factory, and so on have to be decided on. Then, the distributions of exposure amount (concentration) for the group have to be estimated. There are three ways to estimate the exposure distribution: (1) based on the measurement of the concentration in each medium such as air, water, and food; (2) based on the model calculation to estimate the concentration in each medium; and (3) based on biomonitoring, which is a direct method to measure the amount taken up by measuring the chemical compounds in blood or urine. In the case of (1) and (2), the measured/estimated distribution of concentrations is multiplied by the respiration

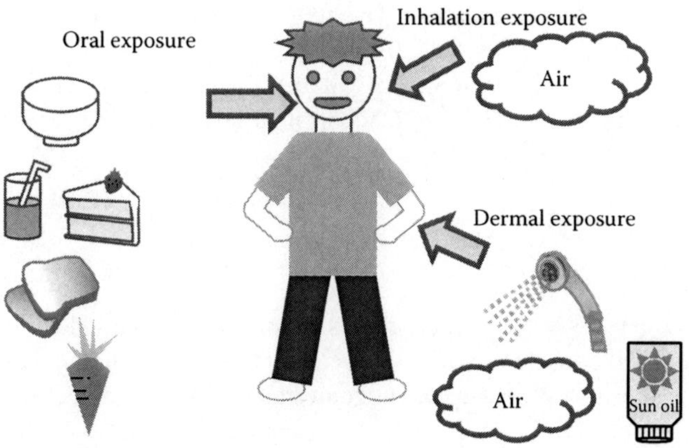

FIGURE 12.8 Three exposure routes.

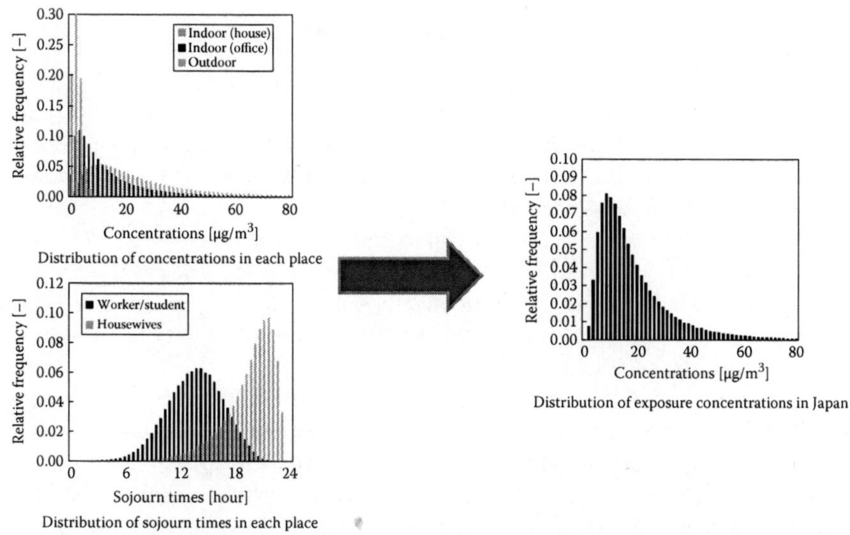

FIGURE 12.9 Derivation of exposure distribution to acetaldehyde in Japan. Arithmetic mean: 20.8 μg/m³, standard deviation: 15μg/m³. Geometric mean: 18.1 μg/m³, geometric standard deviation: 2.7 95%ile: 49.2 μg/m³. (Reprinted from Shinohara N et al. 2007. *Detailed risk assessment document 12*. Acetaldehyde. Tokyo: Maruzen [In Japanese].)

volume or volume not taken up using Monte Carlo simulation. An example of exposure assessment, inhalation exposure to acetaldehyde, is shown in Figure 12.9.

12.3.4 Risk Assessment

In risk assessment, the number of people and populations concerned with health effects and whether countermeasures have to be taken are evaluated and judged by comparing the results of hazard assessment to the results of exposure assessment.

Judgment of risk is often conducted by using the hazard quotient (HQ) or margin of exposure (MOE).

When using the HQ, risk is judged by whether the exposure levels are higher than the standard or guideline value or not, according to the following equation:

$$HQ = \frac{(\text{Exposure level})}{(\text{Standard value or guideline value})}$$

When the HQ is higher than 1, the risk is of concern and more detailed estimation or consideration of countermeasures is needed. On the other hand, when the HQ is below 1, the risk is not of concern and countermeasures are not needed.

When using the MOE, risk is judged by comparing the ratio of exposure levels and NOAEL to the uncertainty factor:

$$MOE = \frac{(\text{NOAEL})}{(\text{Exposure level})}$$

When the MOE is higher than the uncertainty factor, risk is not of concern and countermeasures are not needed. On the other hand, when the MOE is below the uncertainty factor, the risk is of concern and a more detailed estimation or consideration of countermeasures is needed.

The HQ and MOE are basically the same principles of judgment. The HQ is easy to understand in the judgment results, but sometimes might show the results in an arbitrary manner because the uncertainty factor that is used to derive the standard or guideline values is in a black box. On the other hand, the uncertainty factor is not in a black box in judgement with MOE.

The Japanese indoor guidelines of the Ministry of Health, Labor and Welfare were set according to the toxicity assessment assuming only inhalation exposure to indoor air. In the risk assessment, the inhalation exposures in atmospheric and work environments have to be considered for bronchial or pulmonary toxicity. In addition, the other exposure routes such as oral and dermal exposure have to be considered for toxicity to other organs. In the future, these exposures might be considered in revision of the indoor guidelines.

Up to October 2016, the following risk assessments were conducted: primary risk assessment conducted by the National Institute of Technology and Evaluation (NITE) (167 substances) [19], primary environmental risk assessment conducted by the Ministry of the Environment (339 substances) [20], detailed risk assessment conducted by the National Institute of Advanced Industrial Science and Technology (AIST) (25 substances), and risk assessment for manufactured nanomaterials conducted by AIST (3 substances). Azuma [21] conducted a risk assessment of 93 indoor chemical substances and proposed that formaldehyde, acrolein, *p*-dichlorobenzene, benzene, and benzo(a)pyrene have a high risk and require action.

12.4 ENFORCEMENT AND EFFECT OF REGULATIONS

12.4.1 Time Course of Regulations

As described previously, the Ministry of Health, Labor and Welfare (MHLW), Japan, set the indoor guideline values for 13 chemical substances from 1997 to 2002 in response to the spread of the sick-house problem. Subsequently, various regulations were established by various government ministries to improve the indoor environment.

In 1999, the Housing Quality Assurance Act was established by the Ministry of Land, Infrastructure, Transport and Tourism (MLIT) to permit the performance evaluation and display of labeling for housing. Specifically, experts judge the nine performance items, such as stability of residential structures, fire safety, energy saving performance, indoor air quality, and so forth and provide them to consumers. Although whether the builder utilizes this institution is the builder's choice, the institution makes the information on the indoor air quality available to consumers and promotes the provision of better housing. Formaldehyde emission classification; ventilation performance for whole/partial housing; and indoor concentration of volatile chemicals, such as formaldehyde, toluene, xylene, ethylbenzene, and styrene, are displayed as indoor air quality items. Acetaldehyde was added to the list at 2003, but deleted in 2004 because it can be emitted from natural wooden materials.

The Building Act was revised by MLIT in 2003 to add three items for the improvement of indoor air quality: (1) convention on prohibition of chlorpyrifos; (2) restriction on the use of formaldehyde including building materials; and (3) mandatory setting of a 24-hour automatic ventilation system. As chlorpyrifos, which is used under floors for termite control, has strong neurotoxicity, its usage was prohibited for newly built housing. Building materials were categorized into four classes, F☆, F☆☆, F☆☆☆, and F☆☆☆☆, according to formaldehyde emission rates and were limited to use depending on the emission class (Table 12.3) [22]. A 24-hour automatic ventilation system, which can ventilate more than 0.5/h of air exchange rates, has to be established in newly built housing. Exchange rates of 0.5/h of air mean that half of the room air can be exchanged with the outdoor air.

TABLE 12.3
Categorization and Limitation of Wooden Material Based on the Emission Rates of Formaldehyde in the Revised Building Code

Emission Rate	Emission Rate (Dessicator) [mg/L]			
(Chamber) [μg/m^2·h]	Mean	Max.	Category	Regulation
<5	<0.3	<0.4	F☆☆☆☆	No limitation
5–20	<0.5	<0.7	F☆☆☆	Restrictions the usage
20–120	<1.5	<2.1	F☆☆	area
>120	<5.0	<7.0	F☆	Banning the use

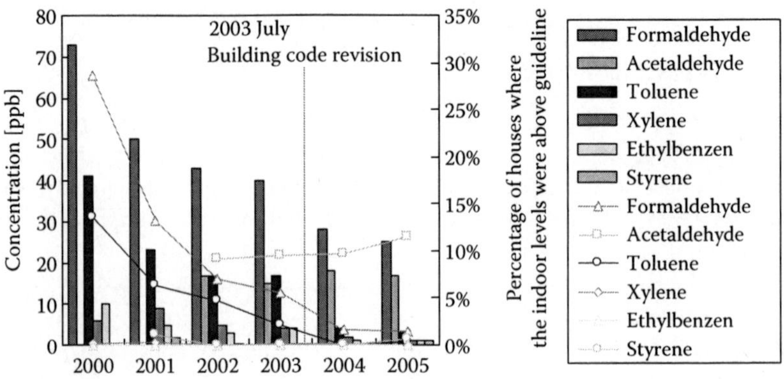

FIGURE 12.10 Indoor concentrations in newly built residential houses surveyed by the Ministry of Land, Infrastructure, Transport and Tourism. Bar shows the indoor concentrations (*left vertical axis*) and the line shows the excess ratio of the guideline value (*right vertical axis*).

Standards of school environmental health were revised by the Ministry of Education, Culture, Sports, Science and Technology in 2002 and 2004 to take action against the sick-school problems. Specifically, the standard demands the measurement of indoor concentrations of formaldehyde and toluene in the classroom every year and when new equipment is delivered. If the indoor concentration exceeds the guideline value, enforcement of ventilation and reduction of the emission will be assumed. If necessary, xylene, ethylbenzene, styrene, and *p*-dichlorobenzene have to be measured.

12.4.2 Impact of Regulations

Although F☆☆ and F☆☆☆ can be used for limited areas in the revised Building Code, F☆☆☆☆ has been used in most houses built after the revision of the Building Code.

MLIT conducted a survey of indoor concentrations of formaldehyde, acetaldehyde, toluene, ethylbenzene, *m/p*-xylene, and styrene in newly built residential houses every year from 2000 to 2005. Except for acetaldehyde, the indoor concentrations of these chemicals were dramatically decreased after 2000 and there are few houses in which indoor concentrations of these chemicals exceed the guideline value of 2005 (Figure 12.10) [23].

ENDNOTES

1. Shinohara N, Naya M, Naka Nishi J. 2007. *Detailed risk assessment document 12.* Acetaldehyde. Tokyo: Maruzen. (In Japanese)
2. Azuma M, Hikita Y. 2002. A study on the indoor air pollution due to the wood preservative: Survey on the under floor and indoor concentrations in newly built wooden house. In *Proceedings of 2002 annual meeting of Japan Society of Home Economics* 54:244. (In Japanese)

3. JIS A 5901. 2004. Straw TATAMIDOKO and straw sandwich TATAMIDOKO. [Japanese]
4. Ministry of Health, Labor and Welfare. 2002. Executive summary of report on judgment standardization of separation of smoking effect. (In Japanese)
5. Onuki A, Saito I, Tada T et al. 2008. Measurements of chemicals related to tobacco smoke in smoking rooms and non-smoking areas. *Indoor Environ* 14(1): 43–50. (In Japanese)
6. Shinohara N, Tokumura M, Kazama M et al. 2014. Indoor air quality and thermal comfort in temporary houses occupied after the Great East Japan Earthquake. *Indoor Air* 24: 425–437.
7. Wechsler CJ, Hodgson AT, Wooley JD. 1992. Indoor chemistry-ozone, volatile organic compounds, and carpets. *Environmental Science & Technology* 26: 2371–2377.
8. Ishizuka Y, Tokumura M, Mizukoshi A et al. 2010. Measurement of secondary products during oxidation reactions of terpenes and ozone based on the PTR-MS analysis: Effects of coexistent carbonyl compounds. *Int J Environ Res Public Health* 7: 3853–3870.
9. Weschler CJ, Shields HC. 2000. The influence of ventilation on reactions among indoor pollutants: Modeling and experimental observations. *Indoor Air* 10: 92–100.
10. Nozaki A, Iikura K, Ikeda K et al. 2002. A study on indoor air pollution by ozone (Part 3): Indoor ozone source and prediction of ozone concentration. In Proceedings of annual meeting of Architectural Institute of Japan, Hokuriku: 977–978. (In Japanese)
11. Follin T. 1996. Measurings during airing out pollutions from concrete slabs. In *Indoor Air'96 Proceedings* 3: 65–70. Nagoya (Japan).
12. World Health Organization. 1983. Indoor air pollutants: Exposure and health effects. EURO reports and studies No. 78, WHO Region Office for Europe, Copenhagen.
13. Kobayashi S, Takeuchi S, Kojima H et al. 2010. Indoor air pollution in a newly constructed elementary school caused by 1-methyl-2-pyrrolidone and Texanol emitted from water-based paints. *Indoor Environ* 13(1): 39–54. (In Japanese)
14. Colborn T. 1997. *Our Stolen Future: Are We Threatening Our Fertility, Intelligence, and Survival?* New York: Plume.
15. Ministry of Health, Labor and Welfare. 2009. 2008 Population Survey Report. (In Japanese)
16. Ministry of Health, Labour and Welfare, Japan. 1999. Survey on volatile organic compounds in residential environment. (In Japanese). http://wwwl.mhlw.go.jp/houdou/1112/h1214-1_13html.
17. National Institute of Advanced Industrial Science and Technology. 2007. Final Report on Comprehensive Chemical Substance Assessment and Management Program Funded by NEDO "Risk Assessment, Development of Risk Assessment Methods and Analyses of Risk Reduction Effects of Risk Management Measures."
18. Ministry of Health Labour, and Welfare, Japan. 1997–2002. Interium report of panel on sick house (indoor air pollution) problem, Environtment Health Bureau, Planning Division: Tokyo.
19. National Institute of Technology and Evaluation. 2016. Chemical substances hazard assessment report, initial risk assessment report. http://www.nite.go.jp/en/chem/chrip/chrip_search/intSrhSpcLst?_e_trans=&slScNm=CI_02_001.
20. Ministry of Environment. 2016. Initial environment risk assessment of chemical substances. (In Japanese). http://www.enr.go.jp/chem/risk/.
21. Azuma K. 2007. *A study on framework for risk assessment and management of indoor air pollutants in residential environment.* Doctoral dissertation, Kyoto University. (In Japanese). http://repository.kulib.kyoto-u.ac.jp/dspace/bitstream/2433/123494/1/D_Azuma_Kenichi.pdf.

22. Ministry of Land Infrastructure, and Transport, Japan. 2003. Notice No. 1113 of the ministry of land infrastructure and transport, Japan Tokyo 2003.
23. Center for Housing Renovation and Dispute Settlement Support. 2005. The survey on the indoor concentrations of volatile chemicals in newly built houses. The Center for Housing Renovation and Dispute Settlement Support, Tokyo, Japan. (In Japanese). http://www.chord.or.jp/tokei/index.html.

13 Methods for Measurement of Indoor Pollution

Atsushi Mizukoshi, PhD, Assistant Professor
Department of Environmental Medicine and Behavioral Science, Kindai University Faculty of Medicine

Indoor pollutants emitted from building materials and consumer products are diffused to indoor spaces and inhaled by residents. At each stage, there are measurement methods for factors such as emission rates from building material, indoor concentration, and personal exposure concentration.

13.1 METHODS FOR MEASUREMENT OF INDOOR CONCENTRATION

To check whether there are any adverse health effects of pollutants in indoor air, it is necessary first to measure the indoor concentration of pollutants. Usually, to know the composition and concentrations of pollutants, the pollutants are collected in an adsorbent and accurate analysis is conducted using analytical instruments. This method is called integration measurement. The concentration obtained by integration measurement indicates an average concentration in the sampling period. On the other hand, there are real-time monitors that can measure total concentrations of pollutants instantly though the composition of pollutants cannot be known. The measurement method by such monitors is called instant measurement. The instant measurement method has the advantage that concentrations can be known on site. Moreover, because the pollutant concentration can be monitored continuously, the change in concentration can be known. Devices that can measure both composition and concentrations of indoor pollutants instantly are rare at present. Therefore, it is necessary to choose integration measurement or instant measurement, based on the purpose of the measurement.

13.1.1 Integration Measurement

Because the sampling and the analysis are conducted separately in the case of integration measurement, accurate analysis (e.g., the separation of chemicals using an analytical column and qualitative and quantitative analysis of each compound by standard compounds) can be conducted in the laboratory. Therefore, the integration measurement method is called the accurate measurement method. The integration measurement method is used to measure the highest indoor concentration under the condition that all windows are shut or to measure the averaged indoor concentration under conditions of daily life. Integration measurement methods are divided into active and passive methods based on the sampling method.

13.1.1.1 Active Method

The active method includes active sampling of chemical compounds by pulling ambient air by a pump through a sampler filled with adsorbent [1]. The chemical compounds adsorbed on adsorbent are eluted by solvent or desorbed thermally and then introduced into an analytical instrument such as high-pressure liquid chromatograph (HPLC) or gas chromatography–mass spectrometer (GC–MS). As HPLC or GC–MS can separate a chemical component using an analytical column, the composition and concentrations of a chemical compounds can be known.

Because ambient air volume sampled is measured accurately by the active method, the concentration can be calculated by dividing the amount of chemical compounds by the sampling volume. However, because a pump and power supply are required, the active method is complicated compared with the passive method.

The general sampling time for measuring an indoor concentration is 30 minutes. It is suitable for measuring the highest concentration in new residential housing.

13.1.1.2 Passive Method

The passive method includes the passive sampling of chemical compounds by exposing a sampler containing adsorbent to ambient air. The chemical compounds in ambient air are collected to the adsorbent by molecular diffusion based on Fick's law. Because a pump and power supply are unnecessary, it is convenient as compared with the active method. However, to calculate concentrations, the uptake rate of each chemical compound is needed beforehand. An uptake rate is a value corresponding to the airflow rate of the pump in the active method, and is obtained by a simultaneous measurement with both the active and passive methods. Generally, because uptake rates are smaller than the flow rates of pumps, the passive method is suitable for sampling over a period of time. The general sampling time is 24 hours. The passive method is effective in a measurement of the concentrations under conditions of daily life.

13.1.1.3 Subject Compounds, Carbonyl Compounds, VOCs, TVOCs, SVOCs

Typical indoor pollutants are carbonyl compounds such as formaldehyde and acetaldehyde, volatile organic compounds (VOCs) such as toluene, semivolatile organic compounds (SVOCs) such as phthalate and organophosphate ester, and so on. It is necessary to choose a sampler and an analytical instrument according to the pollutants.

13.1.1.3.1 Carbonyl Compounds

Carbonyl compounds (formaldehyde, acetaldehyde, acetone, butylaldehyde, etc.) are collected into a sampler filled with silica gel impregnated with 2,4-dinitrophenyl hydrazine (DNPH). The carbonyl compounds are derivatized by the reaction of the carboxyl group and DNPH (Figure 13.1) and adsorbed chemically. The derivatized

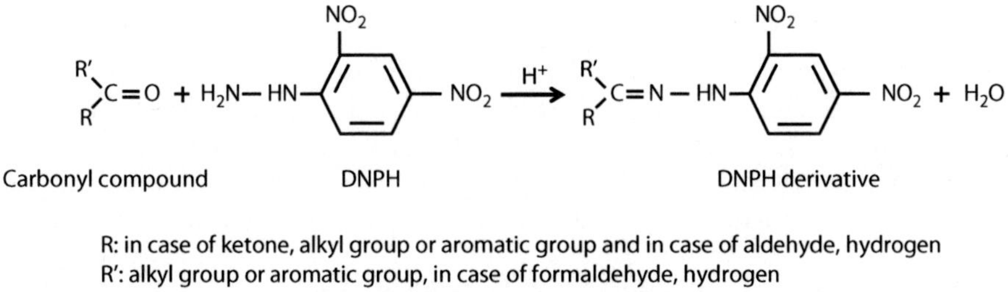

FIGURE 13.1 Reaction of carbonyl compounds and DNPH.

carbonyl compounds can be extracted by solvent (acetonitrile) and quantified by measuring the absorbance at a wavelength of 360 nm by HPLC.

13.1.1.3.2 VOCs and TVOCs

VOCs are collected on active carbon or resin adsorbent (Tenax® TA). VOCs collected on active carbon are extracted by a solvent (carbon disulfide) and analyzed by GC–MS. VOCs collected on resin adsorbent are desorbed thermally and analyzed by GC–MS. Because various VOCs exist indoors and it is difficult to identify and quantify all of them, the concept of total volatile organic compounds (TVOCs) to figure out VOCs in a total was proposed. To calculate TVOC concentration, areas of the peaks during the retention time from hexane to hexadecane on the GC are all integrated. Then the summation of the total area is converted to concentration by using a toluene calibration curve (toluene conversion). The Ministry of Health, Labour and Welfare Japan defined the TVOC as the summation of VOC_{id} and VOC_{un}. VOC_{id} is the sum of concentration of the compounds that can be identified by using standard compounds. VOC_{un} is the sum of the concentration of the compounds that cannot be identified. The concentration of VOC_{un} is calculated by toluene conversion. In the passive method, because the uptake rates of all the VOCs included in the TVOCs cannot be known, TVOC concentration cannot be calculated by the aforementioned method.

13.1.1.3.3 SVOCs

Phthalate compounds such as di-*n*-butyl phthalate (DBP) and di(2-ethylhexyl)phthalate (DEHP) are collected on an adsorbent (carbon adsorbent or styrene-divinylbenzene copolymer, etc.), extracted by solvents (acetone or in the case of carbon adsorbent, dichloromethane), and analyzed by GC–MS after concentrating by blowing nitrogen gas. Because phthalate can be easily affected by environmental contamination, it is important to pay attention to the chemistry apparatus and experimental procedure and reduce contamination as much as possible.

Chlorpyrifos is collected on an adsorbent (styrene-divinylbenzene copolymer or octadecyl-silylated silica gels), extracted by solvent (acetone), and analyzed by GC–MS after concentrating by blowing nitrogen gas. GC–flame photometric detection (FPD) can also be used. In new residential housing, sampling for 2 hours at a flow rate of 10 L/min is necessary to collect a sufficient amount to compare with a guideline value (see Chapter 12).

13.1.2 Instant Measurement

In recent years, real-time monitors that can measure the concentration of indoor pollutants instantly have been developed. By this instant measurement method using these real-time monitors, the pollutant concentration and its temporal change can be known on site. Because an accurate analysis technique is unnecessary, the instant measurement method using a monitor can be called the simplified measurement method.

13.1.2.1 Subject Compounds, TVOCs, Formaldehyde

Some real-time monitors are equipped with a semiconductor sensor or photoionization detector (PID). Because these monitors do not have a separation feature for chemical composition, they measure the TVOCs. Mølhave et al. introduced real-time monitors of TVOC [2] as direct-reading instruments. On the other hand, there are monitors that can measure specific chemical compounds such as formaldehyde [3].

13.1.2.2 On-Site Method of Measuring Source Intensity and Ventilation Rate

As an application of the instant measurement method, by measuring the temporal change of indoor concentration at non-steady state and fitting the expected concentration change formula to temporal change, on-site source intensity and ventilation rate can be estimated [4].

For example, by solving a differential equation of mass balance on the assumption that the initial indoor concentration is equal to the outdoor concentration, the indoor concentration is indicated as in Equation 13.1.

$$C(t) = C_o + E/F \times (1 - e^{-Ft/V}) \tag{13.1}$$

where

$C(t)$ = indoor concentration at time t (μg/m^3)
C_o = outdoor concentration (μg/m^3)
E = emission rate (μg/h)
F = ventilation rate (m^3/h)
t = time (h)
V = room volume (m^3)

Therefore, first ventilate sufficiently by opening windows, make the indoor concentration equal to the outdoor concentration, and close the windows. Then measure the indoor concentration increase by the instant measurement method. By fitting Equation 13.1 to the concentration increase, the emission rate of pollutants and ventilation rate can be estimated.

13.1.3 Combined Application of Integration Measurement and Instant Measurement

Though it is possible to analyze composition by integration measurement, the measurement value is an averaged concentration during sampling. Meanwhile, we can

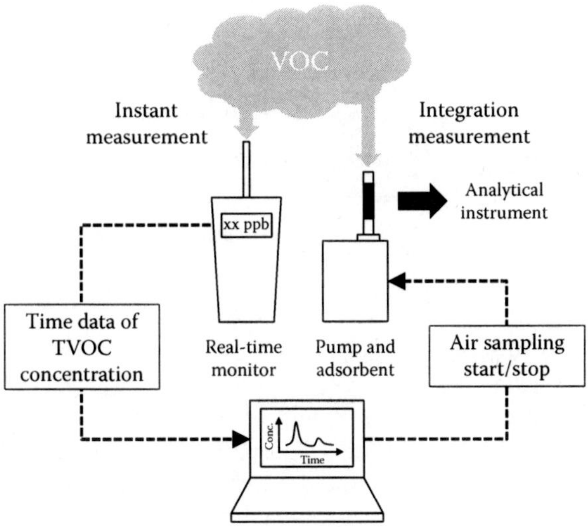

FIGURE 13.2 Peak capture system.

know the change of concentration on site by instant measurement, but the composition is unknown. By combining these methods, it becomes possible to combine the advantages of both methods.

13.1.3.1 Peak Capture Method

Indoor concentrations increase or decrease by residents' activities. To know the composition and concentrations of pollutants at the time at which the concentration increases it is necessary to perform an integration measurement at that time. The concentration change is monitored by instant measurement and if a concentration increase (peak) is predicted, the composition and concentrations of pollutants at the time of concentration are measured by conducting integration measurement. This method is called the peak capture method [5]. A scheme of peak capture system by this method is illustrated in Figure 13.2.

13.2 METHODS OF MEASURING THE EMISSION RATE OF BUILDING MATERIALS

Methods of measuring the emission rate are divided into laboratory methods (a desiccator method and a chamber method) and on-site methods. To confirm the emission rate of building materials before using, the desiccator method and the chamber method can be used. Meanwhile, to explore the emission source of pollutant in an indoor environment, on-site measurement is available.

13.2.1 Desiccator Method

The desiccator method measures the amount of formaldehyde emission amount generated from building materials. The test pieces and crystallizing dish filled with

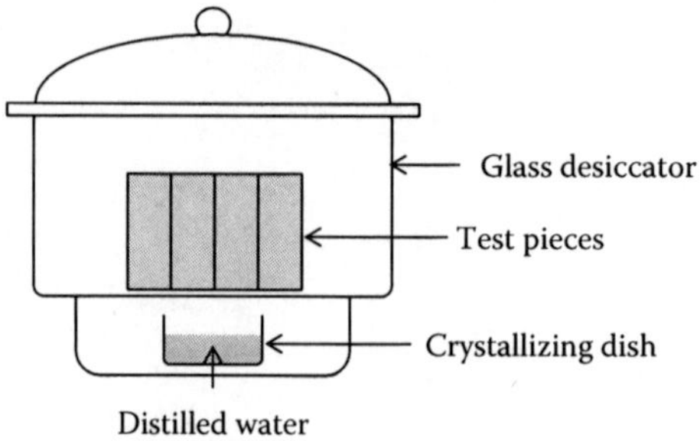

FIGURE 13.3 Glass desiccator method.

distilled water are put in a desiccator and the desiccator is set in a constant temperature reservoir at 20°C for 24 hours (Figure 13.3). Because emitted formaldehyde from test pieces is absorbed into distilled water, the concentration of formaldehyde in distilled water (mg/L) is defined as the amount of formaldehyde emitted. The concentration of formaldehyde in distilled water is measured by the acetylacetone absorptiometric method. This method is provided in JIS A 1460: 2015 [6] by the Japanese Industrial Standards (JIS) and Japan Agricultural Standards (JAS).

13.2.2 Chamber Method

The chamber method measures the emission rate of the chemical compounds emitted from building materials. As shown in Figure 13.4, the test pieces are put into a small chamber (20–1000 L). During the test period, clean air is introduced into the chamber, where the ventilation rate, temperature, and relative humidity are controlled (e.g., the ventilation rate is 0.5/h, temperature is 23°C for ISO and 28°C for JIS, and relative humidity is 50%). The emission rate of the chemical compound per unit area of a test piece can be calculated from the concentration of the chemical compound in the chamber, the ventilation rate, and the surface area of a test piece [7,8].

To measure the emission rate of entire furniture or building materials, the large-chamber (1 to 80 m^3) methods can be used [9,10].

To measure the emission of SVOCs, the microchamber method can be used. Because SVOC adsorbs on the inner wall of the chamber, all the emitted compounds cannot be collected in a sampler. Therefore in the microchamber method, first SVOCs emitted from test pieces are collected in a sampler. Subsequently, test pieces

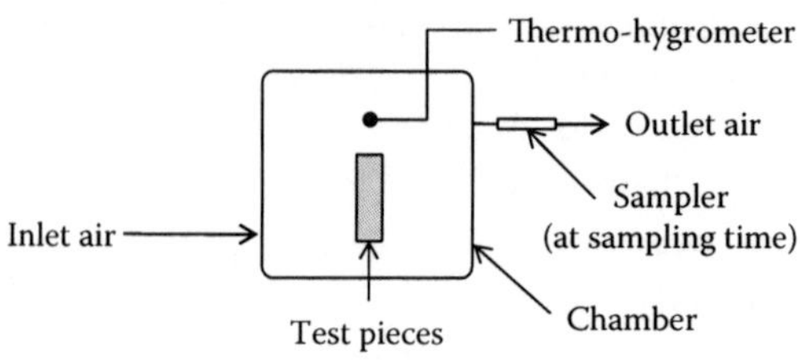

FIGURE 13.4 Chamber method.

are removed from the chamber and SVOC adsorbed in the chamber is desorbed by heating the chamber and collected in the other sampler. The emission rate of SVOC is calculated by adding the amount of emission and desorption collected [11,12].

13.2.3 On-Site Measurement Method

To measure the emission rate of the building materials in a real environment by the desiccator method or the chamber method, it is necessary to make test pieces by cutting building materials and measuring them in a laboratory. On the other hand, because the on-site measurement method can collect and/or measure the emitted chemicals on site, the emission rate can be measured nondestructively. On-site measurement methods are also divided into active and passive methods based on the sampling method. Sampling devices of active method include the field and laboratory emission cell (FLEC). Sampling devices for the passive method include the passive emission colorimetric sensor (PECS), passive flux sensor (PFS), and advanced diffusive sampling emission cell (ADSEC).

13.2.3.1 Field and Laboratory Emission Cell

The FLEC is an instrument that can measure the emission rate from building materials in the field or laboratory. The collection of chemical compounds is based on the active method. The mini test chamber is made from stainless steel (with a volume of 35 mL) and is set on surfaces of building materials. Clean humidified air is flowed into the chamber and chemical compounds are collected in a sampler from the chamber outlet (Figure 13.5) [13]. FLEC can simulate the actual indoor environment because of the flow system.

13.2.3.2 Passive Emission Colorimetric Sensor

PECS, also called the Yanagisawa sensor, measures the amount of formaldehyde emission using an enzyme reaction [14] (Figure 13.6). Formaldehyde dehydrogenase (FALDH), nicotinamide adenine dinucleotide (NAD^+), and coloring reagent are contained in a test sheet in the sensor. By trickling one drop of water into the test sheet of a sensor and placing the sensor on the surface of the material, formaldehyde emitted from material reaches the test sheet. Then formazan is generated by an enzyme

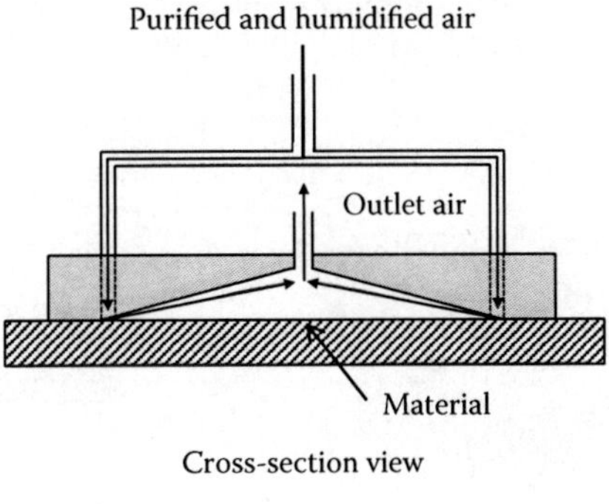

FIGURE 13.5 Field and laboratory emission cell.

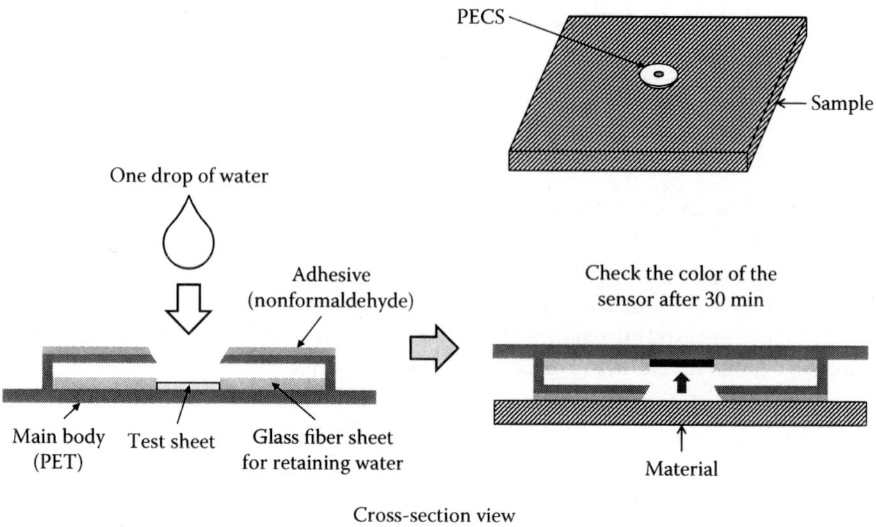

FIGURE 13.6 Passive emission colorimetric sensor (Yanagisawa sensor).

reaction and the color of the test sheet turns red according to the amount of formaldehyde emission after 30 minutes after placement of the sensor. The emission classification can be judged visually from this color strength. Moreover, by measuring the absorbance of the test sheet using a reflectance photometry device, the emission amount (mg/L) or emission rate (μg/m^2·h) can be calculated from a conversion factor.

13.2.3.3 Passive Flux Sampler

A passive flux sampler (PFS) can measure the flux of test materials by a passive method. By putting a small container including absorbent on test materials, chemicals from materials emitted are collected passively on the adsorbent (Figure 13.7).

The flux of chemicals from a material can be calculated by dividing the amount collected by the sampling time and surface area of the adsorbent. The flux of various chemical compounds can be measured by changing the species of adsorbent.

Using PFS, emission rates of carbonyls and VOCs from multiple indoor sources [15] and temperature dependence of phthalate esters from plastic materials [16] are measured.

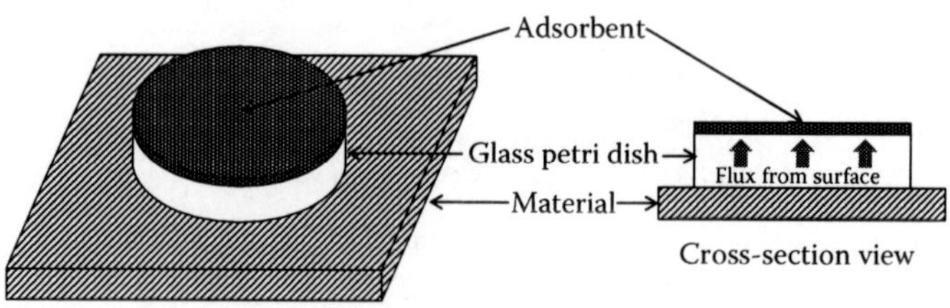

FIGURE 13.7 Passive flux sampler.

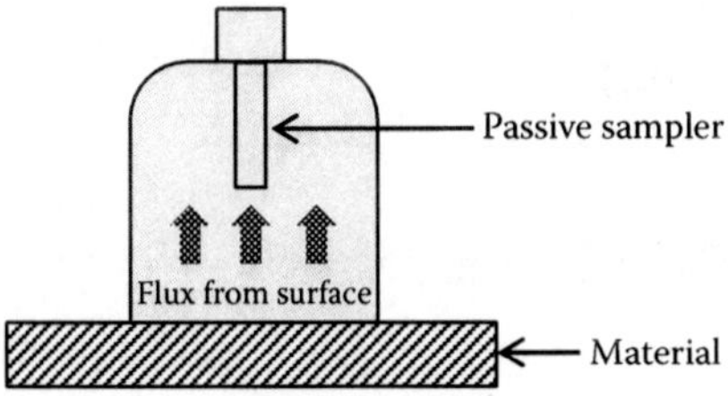

FIGURE 13.8 Advanced diffusive sampling emission cell.

13.2.3.4 Advanced Diffusive Sampling Emission Cell

ADSEC is a device that can measure the emission rate from building materials by the passive method. By putting the small chamber on the material, emitted chemicals are collected in a passive sampler that is set in the chamber (Figure 13.8). The collection time is 24 hours. This device is compliant with JIS A 1903: 2015 [17].

13.3 METHODS OF MEASURING PERSONAL EXPOSURE AMOUNT

To investigate the relationship between chemical exposure and health effects, it is desirable to measure how much the individual is exposed to the chemical compounds. Moreover, the information about personal exposure amount can be used to conduct countermeasures for reduction of chemical compound exposure.

Personal exposure to indoor air pollutants can be known by measuring chemical compounds in ambient air or biological samples such as blood, urine, and exhaled air.

13.3.1 Active Sampling–Passive Sampling Method

The AS–PS method is a personal exposure measurement method that combines active sampling (AS) and passive sampling (PS) [18]. A subject carries a passive sampler and a pump equipped with an active sampler. When a certain chemical exposure is perceived, a pump is started by the subject and the chemical compound is collected to an active sampler. Next, both passive and active samplers are analyzed. From the results of the passive sampler, the averaged personal exposure concentration during the measurement period can be known. Meanwhile, from the result of the active sampler, the concentration at which the subject perceived some kind of exposure can be known. By comparing these values, the chemical compounds with higher concentrations on active sampling are possible compounds to which the subject perceived being exposed.

13.3.1.1 Search for Causative Compound of Chemical Sensitivity

The aforementioned AS–PS method is applicable to explore the causative compounds that provoke the symptoms of chemical sensitivity (CS). That is, a patient carries the sampling device of the AS–PS method and when chemical exposure is perceived and the CS symptom is triggered, he or she conducts active sampling. If the compound's concentration in the active sampler is higher than that in the passive sampler,

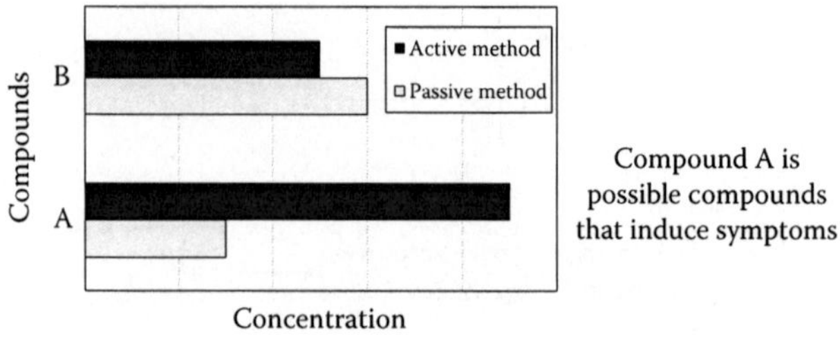

FIGURE 13.9 Result of the AS–PS method.

the compounds have the possibility of triggering CS symptoms. An example of the result of the AS–PS method is shown in Figure 13.9. In this case, it is thought that compound A may have caused the patient's symptoms.

13.3.2 Instant Measurement Method

When investigating the relation between chemical compound exposure and health effects, the measurement must be conducted with the proper time scale to detect each temporal change. For example, instant measurement is needed when a certain symptom appears instantly after exposure to chemical compounds.

13.3.2.1 TVOCs and HRV

As mentioned in Section 13.1.2, there are real-time monitors that can measure instantly. By walking around with this monitor, we can know when, where, and how much we are exposed to VOCs. At the same time, biological effects by VOC can be known by monitoring biological information. As a monitoring device of biological information, a Holter monitor can record electrocardiograms. The function of autonomic nerves can be measured by analyzing heart rate variability (HRV) from the recorded electrocardiogram. When instant measurement of TVOC and monitoring of an electrocardiogram are conducted in parallel, the relationship between TVOC exposure and function of autonomic-nerves in a high time resolution can be investigated [19,20].

ENDNOTES

1. U.S. EPA. 1999. Compendium Method TO-17. Determination of volatile organic compounds in ambient air using active sampling onto sorbent tubes. In *Compendium of methods for the determination of toxic organic compounds in ambient air*, 2nd ed. Washington, DC: U.S. Environmental Protection Agency.
2. Mølhave L, Clausen G, Berglund B et al. 1997. Total volatile organic compounds (TVOC) in indoor air quality investigations. *Indoor Air* 7(4):225–240.
3. Hirst DVL, Gressel MG, Flanders WD. 2011. Short-term monitoring of formaldehyde: Comparison of two direct-reading instruments to a laboratory-based method. *J Occup Environ Hyg* 8(6):357–363.
4. Noguchi M, Mizukoshi A, Yanagisawa Y. 2011. Improvement method of indoor air quality at the new nursery: Evaluation of indoor air quality indicated by TVOC concentration. *AIJ Journal Technol Design* 17(36):577–582.

5. Oka K, Iizuka A, Inoue Y et al. 2010. Development of a combined real time monitoring and integration analysis system for volatile organic compounds (VOCs). *Int J Environ Res Public Health* 7(12):4100–4110.
6. JIS A 1460: 2015. Determination of the emission of formaldehyde from building boards: Desiccator method.
7. ISO 16000-9: 2006. Indoor air—Part 9: Indoor air. Determination of the emission of volatile organic compounds from building products and furnishing: Emission test chamber method.
8. JIS A 1901: 2015. Determination of the emission of volatile organic compounds and aldehydes by building products: Small chamber method.
9. JIS A 1911: 2015. Determination of the emission of formaldehyde by building materials and building related products: Large chamber method.
10. JIS A 1912: 2015. Determination of the emission of volatile organic compounds and aldehydes without formaldehyde by building materials and building related products: Large chamber method.
11. ISO 16000-25: 2011. Indoor air—Part 25: Determination of the emission of semi-volatile organic compounds by building products: Micro-chamber method.
12. JIS A 1904: 2015. Determination of the emission of semi volatile organic compounds by building products: Microchamber method.
13. ISO 16000-10:2006. Indoor air—Part 10: Determination of the emission of volatile organic compounds from building products and furnishing: Emission test cell method.
14. Shinohara N, Kajiwara T, Ohnishi M et al. 2008. Passive emission colorimetric sensor (PECS) for measuring emission rates of formaldehyde based on an enzymatic reaction and reflectance photometry. *Environ Sci Technol* 42(12):4472–4477.
15. Shinohara N, Kai Y, Mizukoshi A et al. 2009. On-site passive flux sampler measurement of emission rates of carbonyls and VOCs from multiple indoor sources. *Build Environ* 44(5):859–863.
16. Fujii M, Shinohara N, Lim A et al. 2003. A study on emission of phthalate esters from plastic materials using a passive flux sampler. *Atmos Environ* 37(39–40):5495–5504.
17. JIS A 1903: 2015. Determination of the emission of volatile organic compounds (VOC) by building products: Passive method.
18. Shinohara N, Mizukoshi A, Yanagisawa Y. 2004. Identification of responsible volatile chemicals that induce hypersensitive reactions to multiple chemical sensitivity patients. *J Expos Anal Environ Epidemiol* 14(1):84–91.
19. Mizukoshi A, Kumagai K, Yamamoto N et al. 2010. A novel methodology to evaluate health impacts caused by VOC exposures using real-time VOC and Holter monitors. *Int J Environ Res Public Health* 7(12):4127–4138.
20. Mizukoshi A, Kumagai K, Yamamoto N et al. 2015. In-situ real-time monitoring of volatile organic compound exposure and heart rate variability for patients with multiple chemical sensitivity. *Int J Environ Res Public Health* 12(10):12446–12465.

14 The Current Situation and Shift in Approaches to Indoor Air Pollution

Miyuki Noguchi, PhD, Research Scientist
Department of Materials and Life Sciences,
Seikei University

Since indoor air has been polluted by chemical compounds, and the resulting adverse health effects have come to be known, various actions have been taken, including passage of laws and regulations. Corporate efforts and social initiatives such as energy savings have also affected the indoor air pollution situation. In this chapter, an administration strategy with laws, regulations, and so forth and their sequelae are explained. In addition, the attitude and perspectives that are needed from now on for the prevention and abatement of health effects such as sick-house syndrome (SHS) are described.

14.1 FRAMEWORK OF LAWS AND REGULATIONS

The following are the main laws and regulations executed in Japan.

14.1.1 Housing Quality Assurance Act (See Table 14.1)

A law specifying quality guidelines for residences and other buildings was enforced by the Ministry of Land, Infrastructure, Transport and Tourism in 2000. In this regulation, information about the existence of the countermeasures against formaldehyde, ventilation systems, and concentration of chemical compounds is required in the housing quality assessment program.

14.1.2 School Health and Safety Act (See Table 14.2)

The standard for school environmental sanitation was revised by the Ministry of Education, Culture Sports, Science and Technology in 2002. In this standard, regular inspections for formaldehyde, toluene, and other chemicals are required in a school environment.

TABLE 14.1
Housing Quality Assurance Act

Inspection Item	Regulation
The countermeasure against a formaldehyde (building, interior finish and under the roof materials)	Class 1: F☆☆ (JIS.JAS) Class 2: F☆☆☆ (JIS.JAS) Class 3: F☆☆☆☆ (JIS.JAS)
Ventilation system	Mechanical ventilation system (above 0.5/h air change rate)
Concentration of chemical compounds	Formaldehyde (essential) Acetaldehyde, Toluene, Xylene, Ethyl-benzene, Styrene (voluntary)

TABLE 14.2
School Health and Safety Act

Inspection Item	Standard Value
Temperature	Winter: 18°C–20°C (>10°C) Summer: 25°C–28°C (<30°C)
Humidity	30–80%
Carbon dioxide (CO_2)	Below 1500 ppm
Carbon monoxide (CO)	Below CO 10 ppm
Air velocity	Below 0.5 m/s
Suspended particulate matter (SPM)	Below 0.1 mg/m^3
Ventilation rate	Kindergarten and elementary school: above 2.2/h Junior high school: above 3.2/h High school: above 4.4/h
Formaldehyde	Below 100 μg/m^3
Toluene	Below 260 μg/m^3
Xylene	Below 870 μg/m^3
p-Dichlorobenzene	Below 240 μg/m^3

14.1.3 Act on Maintenance of Sanitation in Buildings (See Table 14.3)

The Act on Maintenance of Sanitation in Buildings was revised in 2002. In this revision, measurement of formaldehyde concentration (criteria-of-control value below 0.1 mg/m^3) in indoor air in specific buildings (department store, library, office, etc.) was required in addition to measurement of SPM, CO, and CO_2.

14.1.4 Revised Building Standards Act (See Table 14.4)

In addition to the Building Standard Act aimed at planning the security and health of buildings, a revision to prevent sick-house syndrome (SHS) was introduced in 2003.

TABLE 14.3
Act on Maintenance of Sanitation in Buildings

Inspection Item	Standard Value
SPM	Below 0.15 mg/m^3
CO	Below 10 ppm
CO_2	Below 1000 ppm
Temperature	17°C < Temp. < 28°C
Humidity	40% < Hum < 70%
Air velocity	Below 0.5 m/s
Formaldehyde	Below 0.1 mg/m^3

TABLE 14.4
Revised Building Standards Act

Classification	Symbol	Emission Rate (μg/m^2 · h)	Usage Restriction
Out of restriction	F☆☆☆☆	Below 5	No limit
Third-class materials	F☆☆☆	5–20	Limitation of use area
Second-class materials	F☆☆	20–120	
First-class materials	F☆	Above 120	Banning the use

This revision stipulates restrictions on usage of interior finish wood based on the rate of formaldehyde emission; an obligation to install a mechanical ventilation system; and limitation in the use of chlorpyrifos, which is an insecticide for termite control.

However, in addition to regulations on formaldehyde, labeling of building materials with two or more substances such as total volatile organic compounds (TVOCs) is performed in other countries, and evaluation with multiple substances is desired at home.

In terms of regulation of the ventilation rate, installation of a mechanical ventilation system that can secure 0.5/h or above air change in the living room and 0.3/h or above air change in the other rooms of a residence is required.

14.1.5 Indoor Density Guideline Value (See Table 14.5)

The Ministry of Health, Labor and Welfare provided a guideline value for indoor concentrations of volatile organic compounds (VOCs) that will not affect the occupants' health, even if the occupants take in the compounds over a lifetime. These guideline values have been defined for 13 indoor pollutants from 1997 to 2002 since the national survey of formaldehyde was conducted from 1996. Moreover, the indoor concentration of TVOCs, which include VOCs other than the 13 substances, was also determined as an interim desired value of 400 μg/m^3. However, this value was not determined from the toxicological findings, and was determined at concentrations as low as rationally achievable in indoor air quality standards.

TABLE 14.5
Indoor Concentration Guideline Value

VOC	Standard Value of Indoor Concentration	Setting Year
Formaldehyde	100 μg/m^3	1997
Acetaldehyde	48 μg/m^3	2002
Toluene	260 μg/m^3	2000
Xylene	870 μg/m^3	2000
Ethylbenzene	3800 μg/m^3	2000
Styrene	220 μg/m^3	2000
p-Dichlorobenzene	240 μg/m^3	2000
Tetra decane	330 μg/m^3	2000
Chlorpyrifos	1 μg/m^3	2001
Fenobucarb	33 μg/m^3	2000
Diazinon	0.29 μg/m^3	2002
di-*n*-Butyl phthalate	220 μg/m^3	2001
di-*n*-Ethyl hexyl phthalate	120 μg/m^3	2000
TVOC	400 μg/m^3	2001

14.2 TRENDS AFTER DETERMINATION OF GUIDELINE VALUES

According to the survey of the Ministry of Land, Infrastructure, Transport and Tourism, indoor concentrations of substances covered by the guideline have been decreasing every year. However, the result of 2005 fiscal year showed that the percentage of people who experienced health effects caused by chemical substances did not change while they lived in a new building for 1 year. This result indicates the limitations of the countermeasure against source control, based on the guideline substance, and has suggested that responses are necessary not only to the guideline substances but also to VOCs generated in everyday life.

14.2.1 Advantageous Effect of Formaldehyde Labeling

The Revised Building Standard Act in 2003 restricted the use of building materials based on formaldehyde labeling, and articulated the responsibility of setting mechanical ventilation systems that can realize 0.5/h air change and also respond to formaldehyde emitted from furniture.

14.2.1.1 Formaldehyde Concentration in New Residential Housing

According to the results of the survey by Ministry of Land, Infrastructure, Transport and Tourism from the 2000 to the 2005 fiscal year, the formaldehyde concentration was significantly decreased at new residential housing (after completion for less than 1 year), and the percentage of the residences that exceeded the guideline value was 1.5% (18/1181) in the 2005 fiscal year (Figure 14.1).

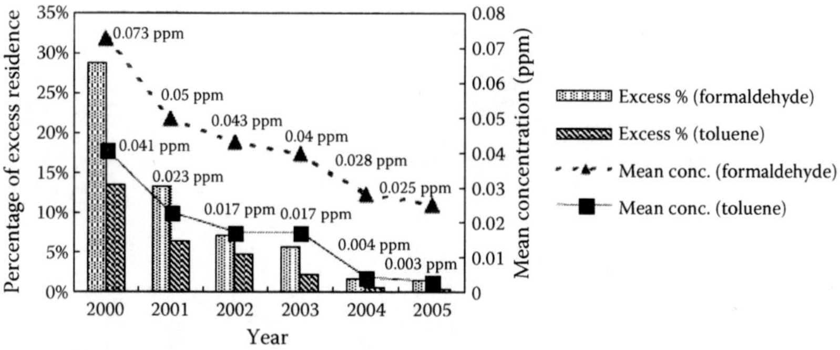

FIGURE 14.1 Survey results of new residential housing.

14.2.2 Composition of VOCs

14.2.2.1 Significant Reduction of Concentrations of Guideline Substances (See Table 14.6)

The concentrations of guideline substances have been reduced significantly since the guideline values were developed, according to the results of the survey of the Ministry of Land, Infrastructure, Transport and Trip.

14.2.2.2 High Concentration Substances (TVOCs, Acetaldehyde, Aliphatic Compounds, etc.)

While the indoor chemical compound is of interest to residents, the safety consciousness regarding chemical compounds spontaneously generated from natural materials is still high. Concentrations of pinene and acetaldehyde that are emitted from wood materials and limonene, which smells like citrus, are often detected at high concentrations indoors because of residents' preferences. Moreover, these compounds are decomposed by air cleaners through oxidation degradation and sometimes generate oxidative substances such as formaldehyde and fine particulates.

On the other hand, concentrations of various VOCs that are generated as a result of residents' activities such as cooking, smoking, dietary habits, and use of chemical products might increase.

14.2.3 Widespread Use of Alternative Substances

Various chemical compounds have been used as alternatives to the guideline substances that were avoided. In many industries, such as the paint and printing industries, water-based solvents and powdery paint have been developed and used. However, it is not always true that water-based solvents are safer than aromatic solvents, including thinner. Diffusion of accurate knowledge to both manufacturers and consumers is desired.

TABLE 14.6
Significant Reduction of Concentration with Guideline Substances

		2000	2001	2002	2003	2004	2005
Formaldehyde	Mean	0.073 ppm	0.050 ppm	0.043 ppm	0.040 ppm	0.028 ppm	0.025 ppm
g.v. = 0.08 ppm	ex.%	809/2815	230/1726	98/1390	84/1491	29/1780	18/1181
		28.7%	13.3%	7.1%	5.6%	1.6%	1.5%
Toluene	Mean	0.041 ppm	0.023 ppm	0.017 ppm	0.017 ppm	0.004 ppm	0.043 ppm
g.v. = 0.07 ppm	ex.%	384/2816	107/1680	67/1390	33/1491	10/1780	4/1181
		13.6%	6.4%	4.8%	2.2%	0.6%	0.3%
Xylene	Mean	0.006 ppm	0.009 ppm	0.005 ppm	0.004 ppm	0.043 ppm	0.043 ppm
g.v. = 0.20 ppm	ex.%	5/2816	5/1680	0/1390	2/1491	4/1780	0/1181
		0.2%	7.1%	N.D.	0.1%	0.2%	N.D.
Ethylbenzene	Mean	0.010 ppm	0.005 ppm	0.003 ppm	0.004 ppm	0.001 ppm	0.001 ppm
g.v. = 0.88 ppm	ex.%	0/2816	0/1680	0/1390	0/1491	0/1780	0/1181
		N.D.	N.D.	N.D.	N.D.	N.D.	N.D.
Styrene	Mean	Unsurveyed	0.002 ppm	0.001 ppm	0.000 ppm	0.000 ppm	0.043 ppm
g.v. = 0.05 ppm	ex.%		18/1680	0/1390	1/1491	1/1780	7/1181
			1.1%	N.D.	0.1%	0.1%	0.6%
Acetaldehyde	Mean	Unsurveyed	Unsurveyed	0.017 ppm	0.015 ppm	0.018 ppm	0.043 ppm
g.v. = 0.03 ppm	ex.%			128/1390	141/1491	172/1780	137/1181
				9.2%	9.5%	9.7%	11.6%

Note: ex.%, exceed%; g.v., guideline value.

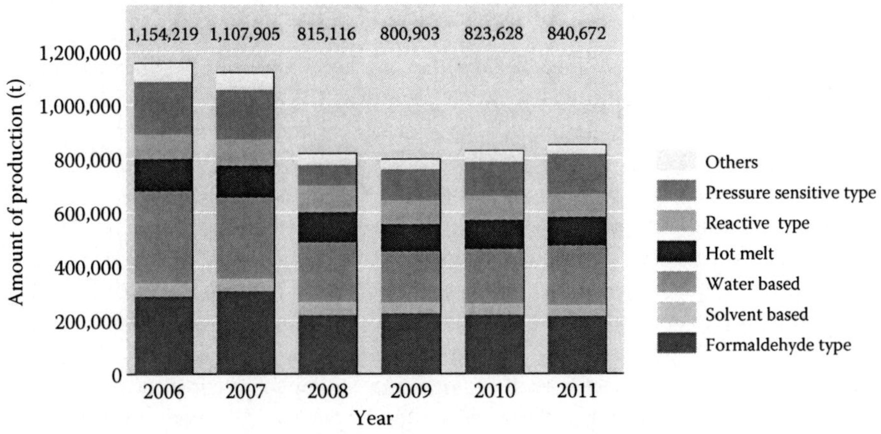

FIGURE 14.2 Trends of production arranged by adhesive type.

14.2.3.1 Low Formaldehyde Emission Adhesives

Medium-density fiberboard, particle board, and plywood used as interior materials are bonded with urea resins, melamine resins, and phenol resins. However, unreacted formaldehyde stays in resins, gradually diffusing and polluting indoor air after construction. Controlled adhesives to decrease the formaldehyde emission by reducing the ratio of formaldehyde or by adding a catcher agent are therefore used. Vinyl acetate emulsions, polyurethanes, and epoxies are also used as formaldehyde-free adhesives, but emission of chemical substances other than formaldehyde may occur, as shown in Figure 14.2.

14.2.3.2 Water-Based Paint and Ink

Low VOC emission paints and inks are being developed and introduced to reduce the emissions from paint and ink. Among these, in terms of water-based paint and ink, many products that use alcohol type solvents are developed and marketed. However, further research and development is still required owing to improvement of workability, quality of products, and so on.

14.2.3.3 Increased Use of Aliphatic Compounds

Aliphatic hydrocarbons are generated mainly from indoor combustion equipment. According to the report of the National Consumer Affairs Center of Japan, when an oil fan heater is used indoors, concentrations of aliphatic compounds such as nonane and decane exceed 100 μg/m^3. Emission of high molecular weight aliphatic compounds, such as decane and undecane, which are contained in the wax used as a finishing agent of flooring materials, is accelerated by use of a floor heater, and might reach high concentrations. In addition, it should be noted that aliphatic compounds are hardly detected even if the concentration is high, owing to the low sensitivity to simplified TVOC monitors that can be used easily for investigation of indoor air.

14.2.4 Temporal Change of Indoor TVOC Concentration

In recently constructed buildings, even if they are new, levels of guideline substances are lower than the indoor guideline value. However, TVOC concentration frequently far exceeds 400 $\mu g/m^3$ as the interim target value. When building materials and so on are sources of VOCs, to decrease TVOC concentration early, performing a bake-out that raises room temperature to accelerate VOC emission or positive ventilation including opening windows is effective. In the case of new buildings, even if a 24-hour mechanical ventilation system was installed according to the Revised Building Standards Act, it is important not to depend solely on it but to use window opening and regional ventilation together to increase the ventilation rate. When a large amount of VOCs would be generated by residents' activities such as using chemical products or dietary habits, appropriate use of regional ventilation can prevent the diffusion of pollutants all over the room, and can reduce for decreasing VOC concentrations.

14.2.5 Trends of Social Recognitions for Indoor Air Pollution: New and Remodeled Houses

Indoor air pollution was considered to be a problem stemming from from carbon dioxide and carbon monoxide for the first time back in the Meiji era. In the 1970s, after adverse health effects caused by formaldehyde were reported in the United States, large-scale investigations were conducted also, and various kinds of regulations such as the guideline values of indoor VOC concentrations were established. It took more than 30 years for SHS to be recognized as a health effect caused by chemical compounds, and its treatment was approved for coverage of medical treatment fees in 2004. Then, the countermeasure of decreasing the concentration of chemical compounds went into full swing, and chemical sensitivity was officially recognized as an illness in 2009. However, though the Ministry of Health, Labour and Welfare specified certain guideline substances, the concentration of total chemical compounds such as TVOCs has not decreased because of the use of alternative substances and numerous natural materials. Moreover, in new buildings in which superinsulated systems for energy conservation were installed, the concentration of chemical compounds has become high as a result.

14.2.5.1 Completion Checks by the Housing Quality Assurance Act

The Housing Quality Assurance Act requires the measurement of formaldehyde (mandatory), acetaldehyde (arbitrary), toluene (arbitrary), xylene (arbitrary), ethylbenzene (arbitrary), and styrene (arbitrary) as a check of air quality when contractors deliver houses to clients. In terms of ventilation, a mechanical ventilation system that can realize 0.5/h air change is required, but measurement of ventilation rate is not.

14.2.5.2 Widespread Indoor Air Pollution with Nonregulated Substances

In investigation of indoor air quality, six of the guideline substances for which measurement is required by the Housing Quality Assurance Act comprise no more than 6% of the TVOC concentration. Figure 14.3 shows survey results of a new nursery constructed with natural materials and containing low VOC emission furniture. Although concentrations of each substance were below the guideline value, TVOC

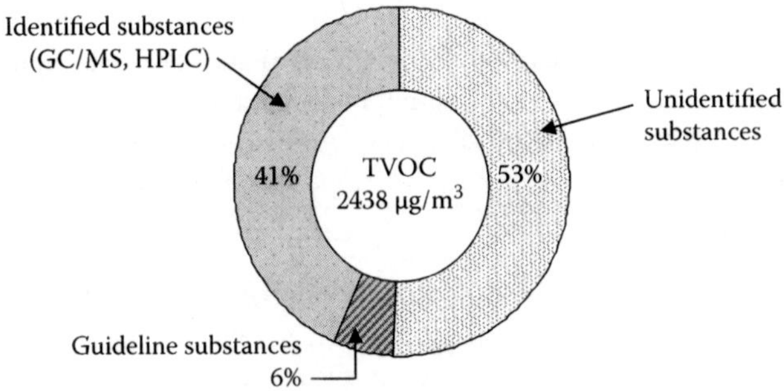

Formaldehyde	9.5 μg m^{-3}	Ethylbenzene	6.4 μg m^{-3}
Acetaldehyde	33 μg m^{-3}	Xylene	8.4 μg m^{-3}
Toluene	17 μg m^{-3}	Styrene	60 μg m^{-3}

FIGURE 14.3 Percentage of VOC in indoor air at a new nursery.

concentration far exceeded 400 μg/m^3, which is the interim target value. It was hard to say that the nursery had indoor air quality that would not cause any adverse health effect. α-Pinene emitted from natural woody materials was a high-concentration compound among identified substances. Significantly, unidentified substances that cannot be determined by instrumental analysis account for more than 50% of TVOCs. Thus it is very inappropriate that indoor air quality is evaluated only based on some substances that were identified as having health effects. In terms of ventilation, even if a mechanical ventilation system was installed and 0.5/h air change was secured on design, unimpeded airflow from the supply port to exhaust port would not be maintained because of the location of the door or ceiling, and so forth, and a room that has an exhaust port acquires negative pressure. Hence the actual ventilation rate should be measured at the time of delivery.

14.2.5.3 Recognition That SHS Was Resolved

According to an actual condition survey of new residential housing conducted by the Ministry of Land, Infrastructure, Transport and Trip, the concentration of guideline substances rarely exceeds the guideline value. Many house builders and consumers believe that SHS is an already solved issue. Therefore, even if a health effect appears after habitation, there is fear that the health effect is not recognized as an issue related to chemical compounds and becomes further exacerbated. As mentioned previously, it is appropriate to use TVOC concentration as an axis of evaluation of indoor air quality at the present stage, as concentrations of VOCs other than guideline substances are still high.

14.2.6 Change in Social Recognition of Indoor Air Pollution: Routine Countermeasures

The most common countermeasures to prevent indoor air pollution are choosing building materials with low emission of chemical substances and installing furniture with only low emission chemical substances. Information regarding formaldehyde is also systematized. When indoor air is still polluted, the countermeasures that remove pollutants by ventilation, as heating instrument makers recommend, are performed. In recent years, use of air cleaners has also increased in response to consumer requests for purer air, including ensuring it is odor free.

14.2.6.1 Air Cleaners

Previous air cleaners used dust-removing filter or active carbon, and so on to remove pollutants. These air cleaners required periodical replacement of filters owing to desorption and diffuse pollutants from an air cleaner after adsorption was saturated. Recently, to put not only VOC but also allergens, molds, and so forth in perspective, air cleaners that use chemical reaction by ozone and free radicals to decompose and remove pollutants have been marketed.

14.2.6.2 Oxidative Decomposition

Recently, air cleaners using ozone, free radicals, photocatalysts, and so on aiming at oxidative decomposition have been developed. However, incomplete oxidative decomposition products are diffused into indoor air because a large amount of energy is needed to decompose VOCs into carbon dioxide and water completely. These products also have an odor due to the oxygen in the molecular structure, and more health effects than undecomposed VOCs.

14.2.6.3 Secondary Pollutants

The substances generated by incomplete oxidative decomposition are named secondary pollutants. They include gaseous incompletely oxidatively decomposed products and fine particles (aerosol) that are part of condensed gaseous secondary products. Health effects caused by fine particles are under investigation; fine particles that are generated through oxidative decomposition are at nano size, and might be carried into the body through the alveolus and vascular system. Besides, α-pinene that is emitted from woody materials and limonene, which used as a deodorant, often reach high concentrations indoors and secondary products are easily generated by oxidative decomposition. It is considered that where concentrations of α-pinene and limonene are high, indoor air quality may worsen further by generation of secondary pollutants from oxidative decomposition of air cleaners.

14.3 REQUIRED OR RECOMMENDED COUNTERMEASURES

Indoor environmental problems lead to other indoor environmental problems, as determined by an understanding of the actual condition, establishment of regulations, and countermeasures. It is necessary to consider countermeasures that also put future environmental issues into the field of view.

14.3.1 Regulation of TVOC Concentration

As described in Section 14.2.5, the guideline substances were determined as target substances thought to be the causes of indoor air pollution resulting in adverse health effects and their concentrations had to be decreased. The concentrations of guideline substances then decreased significantly through the efforts of building materials suppliers and house builders. However, chemical substances have become more diverse and higher in concentration following the use of alternative materials to avoid use of guideline substances, a misunderstanding that natural materials are not harmful, use of a new type of air cleaner, and high air-tightness for energy saving. So, TVOC is also the most appropriate marker that would evaluate indoor problems for the future.

14.3.1.1 Enforcement of the Elimination Period

Most people have thought that it is not necessary to ventilate consciously since 24-hour mechanical ventilation systems have become mandatory. On the other hand, 24-hour mechanical ventilation systems are frequently stopped by residents for energy saving and intolerance of machine noise while sleeping. In new houses, TVOC concentration may be remarkably high just after completion. If ventilation systems are stopped, concentrations of indoor air pollutants would rise instantly. Residents should always be conscious of ventilation, and it is necessary to perform positive ventilation including opening windows in some cases. As a result, emission of pollutants from building materials can be accelerated, and indoor air quality with a low pollutant concentration may be realized in a short period.

14.3.1.2 Information Sharing among the Relevant Parties

Survey studies on alternative materials and generation of secondary compounds from air cleaners are performed by researchers individually. It is an important step of technical development that can respond to future environmental change that house builders and suppliers of building materials and air cleaners share these results and use them for improvement of indoor environment. Academic societies and workshops of ministries and government offices have large roles as information-sharing areas.

14.3.2 Collecting Information from Medical Doctors

Indoor chemical substances have become more diverse, and it has become difficult to identify and link them to health effects individually. It goes without saying that investigations of the health effects of indoor air and efforts to eliminate causative substances are necessary. Toward this purpose, information from medical doctors is indispensable and the creation of systems to distribute the information rapidly to researchers is also important.

14.3.2.1 Selection Method for New Guideline Substances

To select new guideline substances, screening tests using results of animal experiments or biological assays for alternative chemical substances are effective. It is thought that development of screening methods for the estimation of health effects of chemical substances has become an important research endeavor. Further, if causative substances cannot be identified, information calling attention to the present condition is desired.

14.4 FORMULATION OF COUNSELING OR INQUIRY SYSTEMS FOR PATIENTS AND NEW OCCUPANTS

Both house builders and residents recognize that SHS is caused by 13 chemical substances as shown by the guideline levels for indoor concentration. When a health effect appears although the concentration of guideline substances is below the guideline level, it might not be recognized that chemical substances are the causative substances. To reduce such situations, accurate knowledge is indispensable for both suppliers and residents. It is necessary to develop an information-sharing system between researchers and suppliers to make it possible to determine the causes of health effect.

15 Investigation of Indoor Environments and Occupants' Health in Sick Houses

Hiroshi Yoshino, Dr. of Engineering, Professor Emeritus and President-Appointed Extraordinary Professor
Tohoku University
Sachiko Hojo, PhD, Emeritus Professor
Shokei Gakuin University
Rie Takaki, PhD, Associate Professor
Department of Life Design for Safety and Amenity, Faculty of Life Design, Tohoku Institute of Technology

A long-term survey involving collaboration among engineers, medical scientists, and epidemiological and psychological experts was performed for 9 years [1,2]. The study targeted occupants diagnosed with sick-building syndrome caused by chemical substances and their houses. The investigation consisted of measurements of chemical substance concentration and ventilation performance and of questionnaires on occupants' health and the indoor environment.

15.1 FIELD SURVEY ON INDOOR AIR QUALITY, BUILDING PERFORMANCE, AND OCCUPANTS' HEALTH OF 62 SICK HOUSES

15.1.1 Description of the Investigation

15.1.1.1 Investigated Houses

The indoor air qualities of 62 houses in Miyagi prefecture of Japan were investigated in eight summer seasons from 2000 to 2007. The survey period and subject houses are given in Table 15.1. In addition, measurements of the indoor environment were performed in 28 houses for several consecutive or nonconsecutive years from 2001 to 2007. The dwellings surveyed consist of 52 single-family houses

TABLE 15.1
Survey Period and Subject Houses

Terms		Number of Houses	Number of Answers	Response Rate (%)
2000	May–Oct.	23	45	42
2001	June–Oct.	33	137	99
2002	July–Oct.	13	55	93
2003	Aug–Nov.	10	38	83
2004	Aug–Sept.	8	34	92
2005	Aug–Sept.	10	49	96
2006	Aug–Sept.	7	29	100
2007	Aug–Oct.	7	26	100
Total		114	413	84
		(First data: 62)	(First data: 234)	89

Note: 30 houses out of 62 were investigated un/continuously for few years.
(Two times: 20, Three times: 5, Four times: 3, Six times: 2.)

(84%) and 10 multifamily units (16%), and 46 of them are built of wood (79%). There are 35 (56%) new houses that were less than 3 years old at the time when the first survey was performed, and the average age of all is 5 years. Replies to questions about the periods from completion of the building to occupancy were as follows: "occupied before completion," 1 (2%); "occupied in less than 1 week," 36 (58%); "between 1 week and 1 month," 5 (8%); "1 month to 2 months," is 9 (15%); "3 months or more," 6 (10%). Five families (8%) moved to existing apartments. More than half of the occupants started to live in the houses within 1 week after completion. In these houses, indoor air was highly polluted by chemicals. Forty percent (25 out of 64) of the houses were ventilated constantly by mechanical ventilation (except for ventilation under floors) using a supply and exhaust fan (13%) or only an exhaust fan (27%).

15.1.1.2 Investigation of Building Performance and the Indoor Environment

15.1.1.2.1 Contents of the Survey Questionnaire

The questionnaire included building information such as structure, floor plan, building material, as well as life style such as use of insecticides, ways of using the ventilation system, and so on.

15.1.1.2.2 Measurements of Chemical Substance Concentration

Carbonyl compounds (formaldehyde and acetaldehyde) and volatile organic compounds (VOCs; toluene, xylene, *p*-dichlorobenzene, etc., 28 kinds in total) were measured. In some houses, organophosphorus compounds such as chlorpyrifos and others were also measured. Measurement points for carbonyl compounds and VOCs were outdoors and three rooms including a living room and a bedroom where occupants were assumed to be spending a great deal of time, and one more room that was reported to be making people feel sick or having a strong odor. Organophosphorus and other compounds were measured mainly in a Japanese-style room having a tatami mat and otherwise a living room. They were also measured at a point about 1 meter inside the opening of ventilation for a crawl space.

15.1.1.2.3 Measurement of Air-Tightness and Air Change Rate

Air-tightness was measured by the depressurized method using an air-tightness measurement device (Kona Sapporo Co). Air change rate was measured in 14 houses by the constant concentration method. Airflow rate from the supply and exhaust opening was measured in houses with a mechanical ventilation system using an airflow measurement device (Kona Sapporo Co).

15.1.1.3 Survey Questionnaire about Health Conditions

The revised Japanese version [3] of the Quick Environmental Exposure and Sensitivity Inventory (QEESI®) [4] was distributed to occupants. It was used to understand the severity of symptoms and to evaluate the possibility of multiple chemical sensitivity (MCS). Four items except "Masking" consist of 10 questions with scores from 1 to 10, and the total score (0–100) evaluates the level of "Chemical intolerance" and "Symptom severity."

15.1.2 Results

15.1.2.1 Results of Chemical Substance Concentration Measurements

The measurement results of typical indoor chemical substance concentrations are shown in Table 15.2. In the case of formaldehyde, acetaldehyde, and total volatile organic compounds (TVOCs), the ratios of rooms exceeding the guidelines are 64%, 51%, and 60%, respectively. Compared with results of previous surveys in existing normal houses, the excessive ratios for these chemical compounds are considerably higher in the surveyed houses. However, 90% of the houses were built before enforcement of the Amended Building Standard Law of Japan in 2003 to prevent sick houses.

15.1.2.2 Relationship between Formaldehyde Concentration and Air-Tightness Measurement

The relationship between formaldehyde concentration and air-tightness level is shown in Figure 15.1. The figure includes the results measured more than once in different years in the same house. The formaldehyde concentration measured in several rooms was plotted against the same value of air-tightness of a house (total 242 rooms of 82 houses). Air-tightness level of 39 houses (63%), expressed as an equivalent leakage area per floor area, is below the value for an air-tight house of 5.0 cm^2/m^2 applied to the climate zone where the investigated houses are located (Energy Conservation Standard revised in 1999). The formaldehyde concentration is higher in the rooms of air-tight houses with a low air change rate of less than 0.5/h of air.

TABLE 15.2
Measurement Results of Indoor Chemical Substance Concentration

		Sick Houses[a] (2000–2007)				National Surveys for Nomal Houses: (2000)[b]		National Surveys for Nomal Houses: MLIT (2000)[c]		National Surveys for Nomal Houses: MHLW (1997–1998)[d]		
Substance	Unit	Mean	Median	Max.	Over Guideline (%)	Mean	Over Guideline	Mean	Over Guideline	Mean	Over Guideline	Guideline of MHLW
Formaldehyde	μg/m³	137.5	130.9	339.4	64	113.5	32.2%	87.2	27.3%	–	–	100
Acetaldehyde	μg/m³	125.8	110.1	369.7	51	–	–	30.6	9.2%	–	–	48
Toluene	μg/m³	110.2	39.4	1753.3	9	–	–	143.2	12.3%	96.0	–	260
Ethylbenzene	μg/m³	20.1	7.1	489.0	0	–	–	34.7	0.0%	22.1	–	3800
Xylene	μg/m³	26.3	13.8	304.4	0	–	–	21.7	0.1%	36.1	–	870
p-Dichloro benzene	μg/m³	263.2	36.5	8445.4	11	–	–	–	–	125.7	–	240
TVOC	μg/m³	1240.5	511.9	8878.6	60	–	–	–	–	–	–	400

[a] 166 rooms in 62 houses. Maximum value is used from all measurement.
[b] "Survey in Tohoku region in Japan" Objects of study are 59 houses.
[c] "Survey by Ministry of Land, Infranstructure and Transport (2000)" Unit is coverted from "ppm" to "μg/m³" in condition of 25°C. Objects of study are 4368 houses consisting of single-family houses (67%) and one year old or less houses (63%).
[d] "Survey by Ministry of Health, Labour and Welfere." Objects of study are 385 houses.

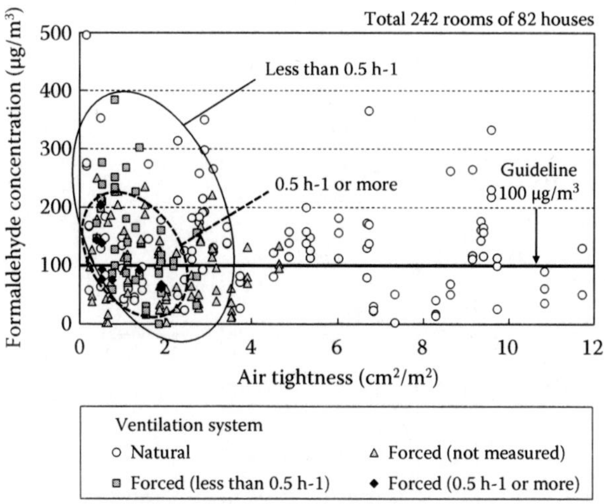

FIGURE 15.1 Relationship between formaldehyde concentration and air-tightness.

15.1.2.3 Relationship between Formaldehyde Concentration and Air Change Rate

Air change rate was measured in 41 rooms of 14 houses out of 62. Figure 15.2 shows the relationship between air change rate of rooms and formaldehyde concentration. In some houses, measurements were performed several times. The figure shows the first measurement results. In only 5 houses out of 14 (36%) was air change rate higher than 0.5/h, which is the minimum value required in principle for new houses regulated by the Amended Building Standard Law of Japan. In 3 houses the air change rate was less than 0.3/h. One house had a value in excess of 1/h. The rooms with lower air change rate such as below 0.5/h tend to show a higher formaldehyde concentration.

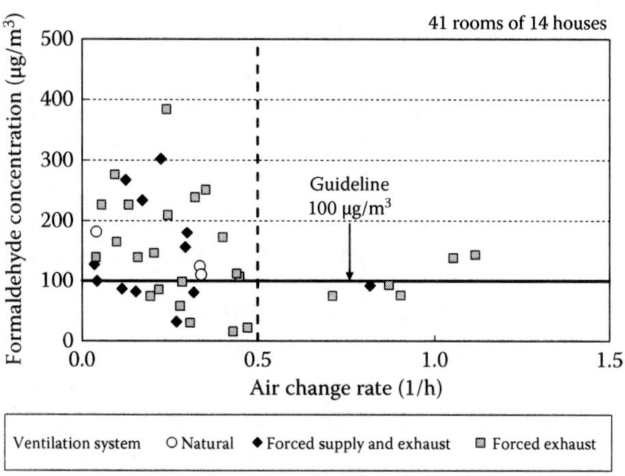

FIGURE 15.2 Relationship between formaldehyde concentration and air change rate.

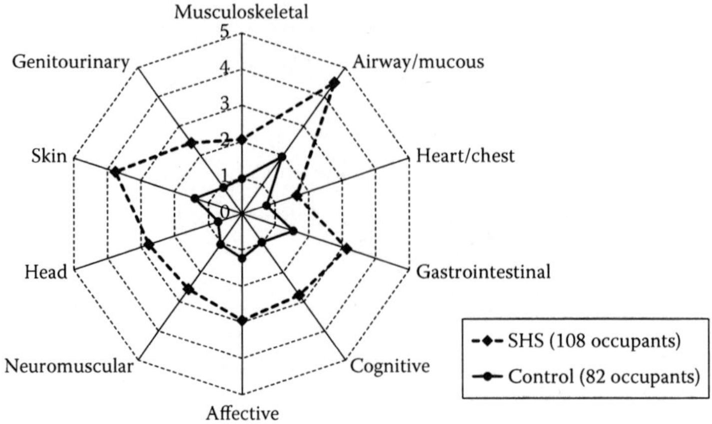

FIGURE 15.3 Questionnaire results (average) regarding 10 subjective symptoms of 105 SHS patients.

15.1.2.4 Sick-House Syndrome Based on the QEESI® Questionnaire

15.1.2.4.1 Classifications of Occupants

A total of 234 occupants answered the QEESI® questionnaire. The results showed that 108 occupants (46.2%) out of 234 were suspected of having sick-house syndrome (SHS), whose symptoms began or became worse less than a year after moving into the new or renovated houses. It was suspected that the symptoms of 44 occupants (18.8%) out of 234 were unrelated to the indoor environment of the houses. The remaining 82 occupants (35%) were found to have no symptoms, and they were categorized as controls.

15.1.2.4.2 SHS Symptoms

Questionnaire results for 10 items regarding subjective symptoms of the SHS (108 occupants) group and the control (82 occupants) group are shown in Figure 15.3. Items include musculoskeletal, airway/mucous membrane, heart/chest-related, gastrointestinal, cognitive, affective, neuromuscular, head-related, skin, and genitourinary symptoms. Higher points mean more serious symptoms. The severity of symptoms was diagnosed by QEESI®. The average of the SHS group was higher than that of the control group for all symptoms. The averages for airway/mucous disorder and skin irritation are especially higher.

15.1.3 Chemical Substance Concentration and Occupants' Symptoms

15.1.3.1 Relationship between Chemical Substance Concentration and SHS

Results of comparative analyses of chemical substance concentration between the SHS group and the control group are shown in Figure 15.4. A *t*-test or Wilcoxon rank sum test was used. All chemical substance concentrations except those of acetaldehyde are higher in the SHS group. The concentrations of toluene, ethylbenzene, xylene, *p*-dichlorobenzene, and TVOCs in the SHS group are higher than in the control group ($p < 0.05$).

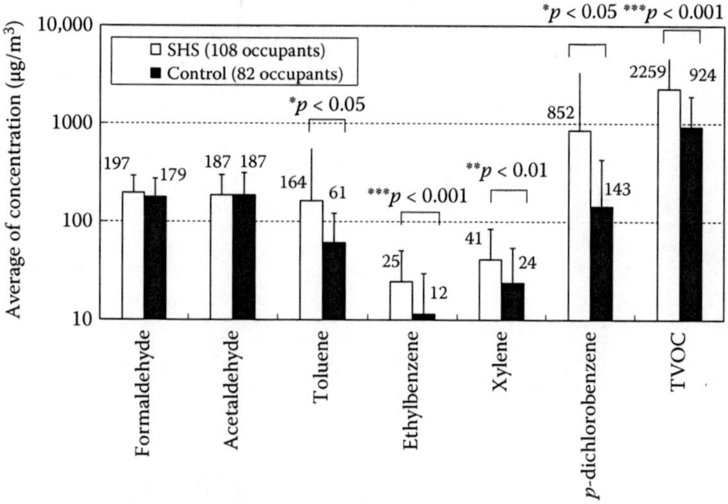

FIGURE 15.4 Comparative analyses of chemical substance concentration.

15.1.3.2 Relationship between Chemical Substance Concentration and Severity of Symptoms

Table 15.3 shows comparative analyses of chemical substance concentration and symptoms between two groups using a *t*-test or Wilcoxon rank sum test. Based on the severity of symptoms diagnosed by QEESI®, occupants were divided into two groups: a group of occupants with a zero score and a group with a score of more than 1. Seven symptoms shown in the table have a significant relationship with the concentration of more than one chemical substance. From these results it is suggested that chemical substance concentration affected occupants' health.

15.1.4 Conclusion

Based on an investigation of 62 houses, it was found that the severity of airway, mucous, and skin symptoms is strong for occupants suspected of having SHS. Many houses had an air change rate of less than 0.5/h even though a mechanical ventilation system had been installed. In houses with high air-tightness performance and low ventilation rate, formaldehyde concentration was rather high. It was confirmed that not only building components such as the wall material and ventilation system but also occupants' behavior such as ventilation operation or use of chemicals in daily life had a strong relationship with chemical substance concentration.

15.2 LONG-TERM OBSERVATIONS MAINLY FOR 30 HOUSES

Long-term observation was performed for 30 houses suspected as sick buildings in Miyagi prefecture in Japan. The study was designed to observe the progress of occupants' health and indoor air quality and also to analyze the relationship between indoor air quality and occupants' behavior and building materials.

TABLE 15.3
Relationship between Chemical Substance Concentration and Symptoms

Symptoms		*N*	Formaldehyde (μg/m³)	Acetaldehyde (μg/m³)	Toluene (μg/m³)	Ethylbenzene (μg/m³)	Xylene (μg/m³)	*p*-Dichloro-benzene (μg/m³)	TVOC (μg/m³)
Airway/mucous	No	55	194	188	81	12**	24**	152*	1007**
	Yes	135	191	187	138	21	37	717	1887
Heart/chest	No	118	192	176	119†	16*	31	270*	1198**
	Yes	72	191	206	124	23	37	1018	2345
Gastrointestinal	No	73	204	195	78†	14**	28	146*	974***
	Yes	117	184	182	149	22	36	811	2051
Cognitive	No	89	203	187	106	16	32	151*	1063**
	Yes	101	182	188	135	21	35	910	2136
Affective	No	89	198	196	100*	16	33	343	1224*
	Yes	101	186	179	140	21	34	736	1990
Neuromuscular	No	97	190	181	128	16	32	129*	1091*
	Yes	93	194	193	114	21	34	991	2190
Skin	No	76	185	192	77†	16	30	348	1302†
	Yes	114	196	184	151	20	35	686	1846
Total	Moderate	113	200	194	100	17	32	290†	1187**
	to intense	77	180	177	154	21	35	946	2299

† $p < 0.1$ *; $p < 0.05$ **; $p < 0.01$ ***; $p < 0.001$.

15.2.1 Investigated Houses

The investigated houses consisted of 25 single-family houses and 5 multifamily units. The investigation was performed two times or more, for a maximum six times un/continuously from 2001 to 2008, from May to September. Twenty-two houses out of 30 were diagnosed as sick houses.

15.2.2 Results of the Investigation

15.2.2.1 Relationship between Chemical Substance Concentration and Building Age

Relationships between building age and formaldehyde concentration as well as TVOC concentration are shown in Figures 15.5 and 15.6, respectively. The same house is connected with a line. Formaldehyde concentration shows a small decrease as the building age increases. The thick line shows houses where a countermeasure for a sick house such as change of ventilation system or replacement of finishing materials was taken in the period of observation. After renovation, the concentration decreased. However, there are two houses with dotted lines, which show an increase only in formaldehyde at the second and final measurements. Occupants of one of them have installed new furniture. The other is a measurement result in a closet without ventilation. TVOC concentration decreases notably as the building age increases. High TVOC concentration was detected in the houses where *p*-dichlorobenzene concentration as shown in parentheses in the figure accounts for most of the TVOC concentration. These houses used moth crystals for clothes.

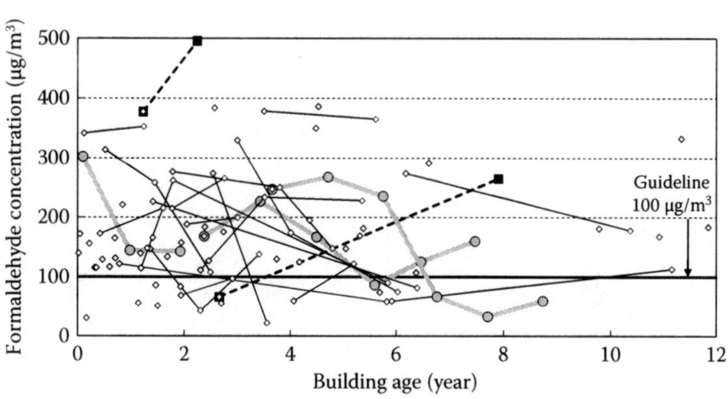

FIGURE 15.5 Relationship between yearly change of formaldehyde concentration and building age (or years after rebuilding).

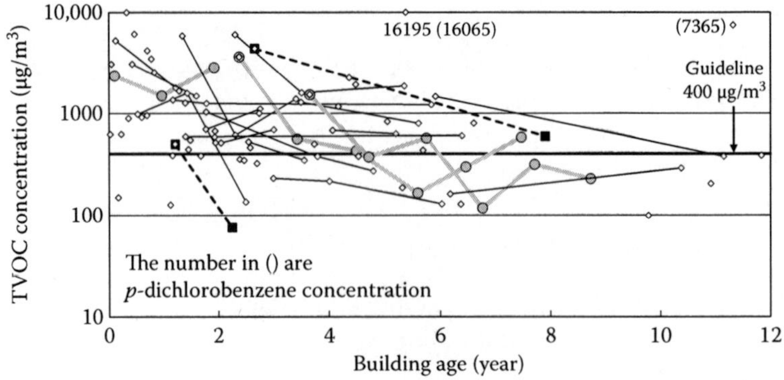

FIGURE 15.6 Relationship between yearly change of TVOC concentration and building age (or years after rebuilding).

15.2.2.2 Yearly Changes of Occupants' Symptoms

Figure 15.7 shows the difference in severity of 10 symptoms between the first and the final QEESI® investigation. The change in symptoms of 56 SHS occupants in 30 houses was examined based on the questionnaire, which has been done several times during the survey period. A decrease in scores, which means recovery from symptoms, was found for five symptoms including head-related, skin, genitourinary, gastrointestinal, and heart/chest-related symptoms. Conversely, an increase was found in affective and neuromuscular symptoms. There was no change in cognitive, musculoskeletal, and airway/mucous membrane symptoms.

Figure 15.8 shows the symptom scores for each occupant at the first and the final questionnaires. Occupants were divided into four groups. "Mild" indicates a group of occupants with a QEESI® symptom score of less than 20 points on both questionnaires. "Recovered" indicates the group of occupants in which the symptom score

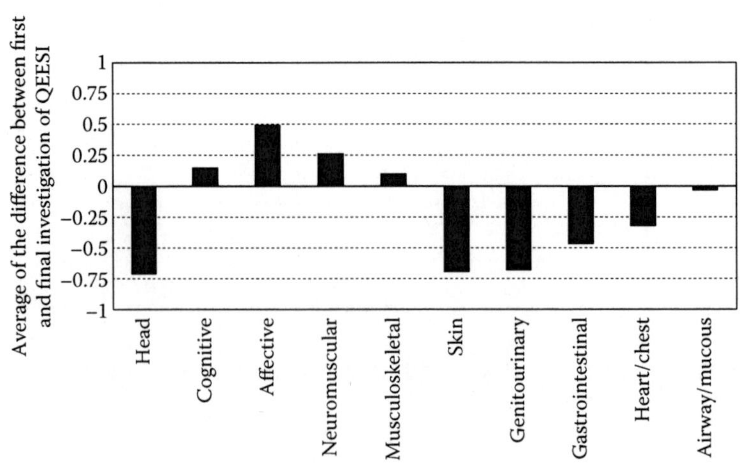

FIGURE 15.7 Differences in severity of symptoms diagnosed by QEESI®.

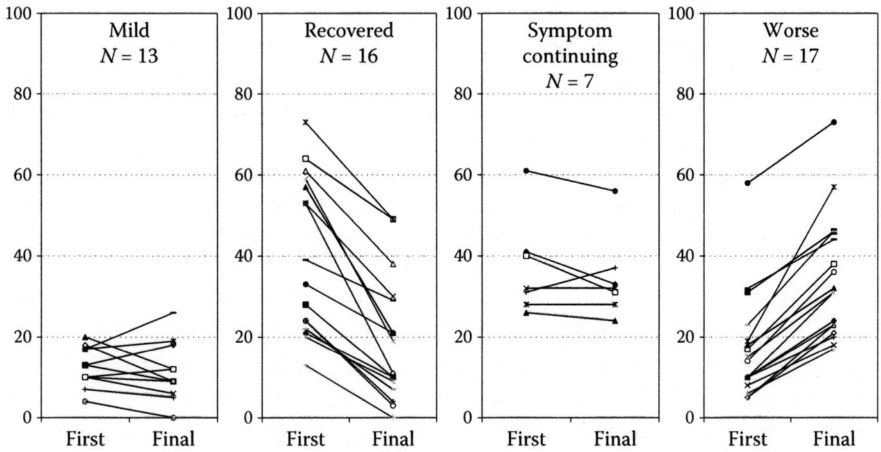

FIGURE 15.8 Yearly changes of occupants' symptom.

decreased by 10 points from the first to the last questionnaire. "Symptom continuing" indicates the group of occupants with a symptom score of more than 20 points on both questionnaires. "Worse" indicates the group of occupants with a symptom score that increased by 10 points from the first to the last questionnaire.

15.2.3 Change of Symptoms and Chemical Substance Concentrations

15.2.3.1 Change of Symptoms in Response to Countermeasures against Mitigation of SHS

Figure 15.9 shows the change of symptoms in 36 sick houses in response to the countermeasures taken for mitigation of SHS. The results came from long-term

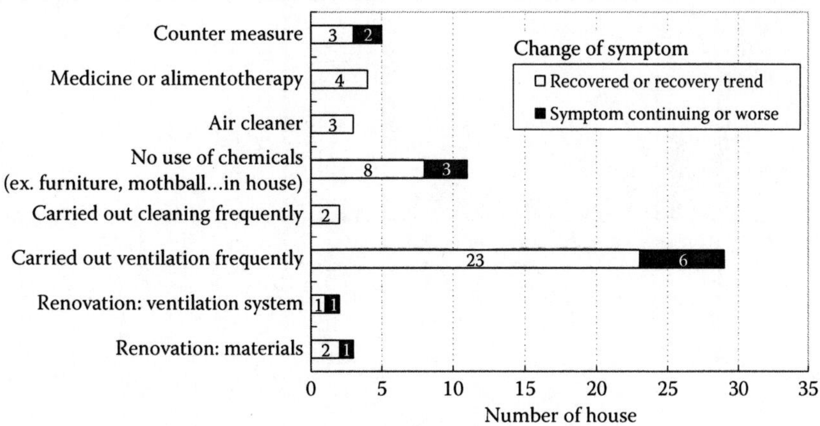

FIGURE 15.9 Result of follow-up study on measurements of chemical substances against health problems and change in symptoms.

observations of 30 houses and follow-up telephone surveys of 18 houses in 2006–2007. The occupants in 36 houses out of 48 were suspected to have SHS. The "Recovery trend" in Figure 15.9 means "recovered from part of symptoms" or "start on a recovery trend." It can be seen that in 29 houses, ventilation was used frequently and in 11 houses, chemical products were not used or products with low chemical content were used for furniture, mothballs, and so on. It was also found that 23 houses out of 29 (ventilation frequently) and 8 houses out of 11 (non-use of chemical products or use of low chemical products) were "Recovered or Recovery trend."

15.2.3.2 Decrease of Chemical Substance Concentration in Response to Countermeasures against Sick Houses

Figure 15.10 shows a decrease of chemical substance concentration between the first measurement and the last one in the case of two time measurements or a decrease between the first measurement and the other one in which the highest value was observed in the case of more than two time measurements for two groups of houses: one has natural ventilation, and the other one has a mechanical ventilation system. A Mann–Whitney test was performed. Figure 15.11 shows the same results for two groups of houses: occupants concern for effects of chemicals on life (e.g., furniture, mothballs), and the other don't concern for them. It was found that from Figure 15.10 that the house with a mechanical ventilation system shows a greater decrease of concentration compared with houses with natural ventilation, especially for formaldehyde, acetaldehyde, TVOCs, aliphatic hydrocarbons, halogenated hydrocarbons, and terpenes. Also, another analysis related to ventilation shows that houses with

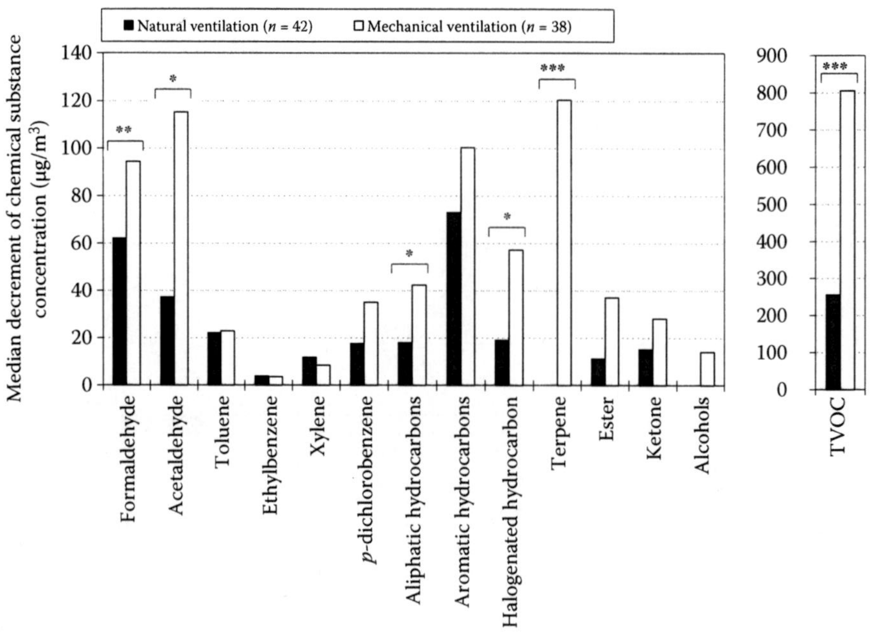

FIGURE 15.10 Decrease of chemical substance concentration based on the type of ventilation.

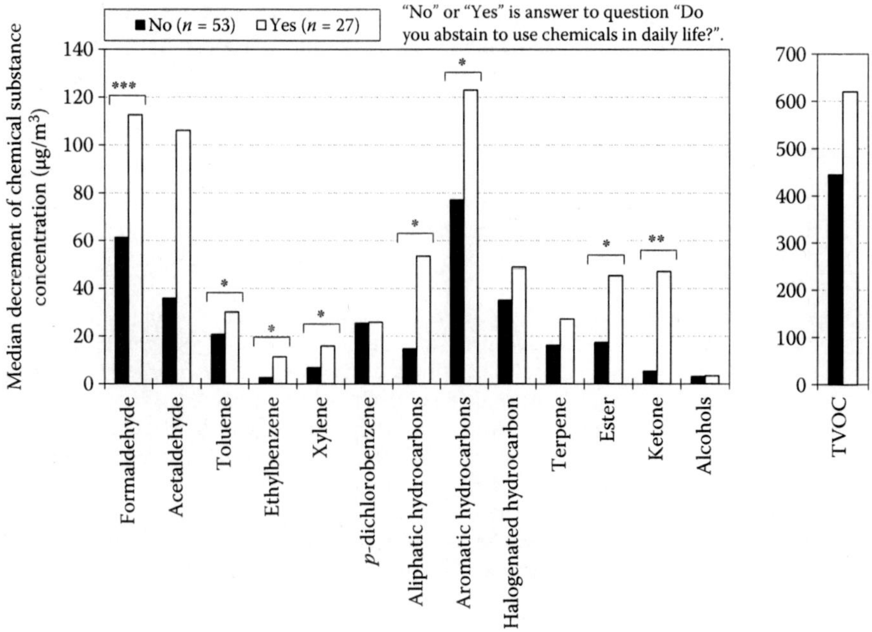

FIGURE 15.11 Decrease of chemical substance concentration based on concern for effects of chemicals on life (e.g., furniture, mothballs).

a higher ventilation rate have a significantly decreased toluene concentration. In houses where occupants ventilated frequently, a decrease was found in toluene, ethyl benzene, xylene, and ketone levels. On the other hand, the decrease in terpene was very large in houses with no regard for ventilation because the group included newly constructed houses where terpene was released significantly from new building materials. Figure 15.11 shows that in houses where occupants avoided using chemicals in daily life, the decrease in concentrations of formaldehyde, toluene, ethyl benzene, xylene, aliphatic hydrocarbons, esters, and ketones is very large.

15.2.4 Conclusions

Based on the nine-year long-term field survey for sick houses, the following results were found:

1. The concentration of chemical substances in general decreased with time, but *p*-dichlorobenzene and formaldehyde levels could increase with the use of mothballs or installation of new furniture.
2. Among 10 symptoms, an increase of the severity score between the first investigation and the last one was found in affective and neuromuscular symptoms.
3. In houses where the severity of symptoms lessened, the concentrations of TVOCs decreased.

4. Frequent ventilation and non-use of products that included chemicals were effective in reduction of chemical substance concentrations as well as in recovery of symptom severity.

ENDNOTES

1. Yoshino H, Nakamura A, Ikeda K, Nozaki A, Kakuta K, Hojo S, Amano K, Ishikawa S. Field survey on IAQ and occupant's health in sick houses. Field survey on 62 houses in Miyagi prefecture. *Journal of Environmental Engineering (Transaction of AIJ)* 74(641):803–809. (In Japanese)
2. Yoshino H, Nakamura A, Ando N, Ikeda K, Nozaki A, Kakuta K, Hojo S, Amano K. Field survey about IAQ and occupant's health of sick house. Long-term field survey about IAQ and occupant's health of 30 house in Miyagi. *Journal of Environmental Engineering (Transaction of AIJ)* 75(654):705–712. (In Japanese)
3. Hojo S, Yoshino H, Kumano H, Kakuta K, Miyata M, Sakabe K, Matsui T, Ikeda K, Nozaki A, Ishikawa S. 2005. Use of QEESI© questionnaire for a screening study in Japan. *Toxicology and Industrial Health* 21:113–124.
4. Miller CS, Prihoda TJ. 1999. The environmental exposure and sensitivity inventory (EESI): A standardized approach for measuring chemical intolerances for research and clinical applications. *Toxicology and Industrial Health* 15:370–385.

16 Round Table Discussion

Yukio Yanagisawa, Chairman,
Professor Emeritus[1] *and Principal*[2]
[1]University of Tokyo
[2]Kaisei Academy Junior and Senior High School

16.1 THE CURRENT SITUATION OF INDOOR AIR POLLUTION

Y. Yanagisawa: We have studied the problems of sick-house syndrome and chemical sensitivity for a long time. I would like to compare the state in early stages of these problems with the present state. What kind of change arose? Dr. Yoshino, please speak from the field of construction.

H. Yoshino: The biggest cause of the early sick-house problem was formaldehyde. After releasing a guideline from the Ministry of Health, Labor and Welfare for the construction industry, concern heightened regarding formaldehyde. The construction industry then came to refrain from use of formaldehyde. Investigations by the Ministry of Land, Infrastructure and Transport showed that the concentration of formaldehyde decreased gradually. This is an outline of early stages of the sick-house problem in the construction industry.

In the following stage, the problem of room air pollution becoming good, with about 13 substances containing formaldehyde appearing in the guideline of the Ministry of Health, Labor and Welfare. On the other hand, when total volatile organic compounds (TVOCs), which do not specify the particular substances, are analyzed, very high concentrations are frequently observed. TVOCs include not only the 13 chemicals but also all VOCs. A value higher than 1000 $\mu g/m^3$ may be observed in the building immediately after construction. I think that new substances other than the 13 substances have been polluting indoor room air recently. The pollution problem caused by these new substances was observed also in the makeshift houses after a great earthquake in East Japan in 2011.

Y. Yanagisawa: Do I hear that TVOC levels are high in a makeshift house?

H. Yoshino: Yes, the makeshift house demonstrates the typical tendency of the latest residence.

Y. Yanagisawa: What kind of change has occurred on the patient side in response to a change of such indoor air pollution chemical substances?

S. Ishikawa: When the concentration of chemical substances such as formaldehyde, toluene and *p*-dichlorobenzene was high, a dose–response relationship existed between the concentration and the symptom. Now, the concentration of chemicals, such as formaldehyde, became low. It seems that the relation between the TVOC concentration and the symptom was realized since the indoor concentration of the 13 guideline substances decreased.

In addition to TVOC concentration, an individual chemical substance such as phthalate ester is important as an indoor contaminant. This substance has attracted attention in the United States. It is reported that caution is required regarding indoor concentrations in schools and hospitals. I think that the method of regulating both an individual substance and TVOC is effective.

Y. Yanagisawa: Dr. Miyata, you are treating patients now. How do you feel about the latest tendency?

M. Miyata: The main condition at the time in early stages of a sick-house problem was the stimulus condition of membranes. The early symptoms were flickering eyes, itchy throat, and headache. These were near poisoning conditions. These days, such an intense symptom is not observed. Instead, before patients understand their illness clearly, they already have sick-house syndrome. Although there were many patients who experienced a disease from new residential housing in the early stages of a sick-house problem, many patients these days are from the rebuilt and remodeled residences. At present, the condition has occurred not only in residences but also in offices. The situation is similar to the sick-building syndrome that occurred in Europe and the United States in the 1980s.

Y. Yanagisawa: If the foregoing is summarized, the latest tendency is not in the situation in which only the specific chemical substance has started the sick-house problem. Various chemical substances are observed in indoor air. It is difficult for a person experiencing sick-house syndrome to recognize that the symptom is due to a specific chemical substance. Although attention needed to be paid to 13 specific substances for which the guideline value was defined, various chemical substances have been involved recently. Moreover, as mentioned, the sick-house problem has occurred not only in residences but also in offices.

The sick-building problem of the Members' Office Building of the House of Councilors that was completed recently is a good example. TVOC levels were high immediately after completion. The most important building of a country is a sick building.

Readers who use this book to study sick-house syndrome and chemical sensitivity observe two kinds of cases: those in the early state in Japan and those in the latest state.

16.2 MEDICAL CORRELATIONS IN JAPAN

S. Ishikawa: I had not done the load examination investigating a dose–response relationship during the early days because specific symptoms appeared after exposure to a given chemical substance in the early stages. For example, eye symptoms caused by formaldehyde were flickering, pain, and tearing. Throat symptoms caused by formaldehyde were acridity and pain. Thus, a simple membrane stimulus symptom clearly had appeared.

If a person spends several hours in a clean room where VOC concentration is kept very low, masking will separate the person from the effects. Masking is a phenomenon in which a person's body adapts itself to the VOC concentration in the air that the person breathes. By adaptation, progression to sick-house syndrome symptoms and a chemical sensitivity can weaken.

Therefore, even if a load examination is done in the state where masking has occurred, I cannot clarify an exact dose–response relationship. If clean room equipment can be ready and the influence of masking can be removed, the exact dose–response relationship for a specific chemical substance will be clear.

However, these days, symptoms have become complicated. For example, symptoms connected also with mental disturbances have been shown. If the proficient doctor conducts an oral consultation carefully, he or she can discern whether the patient is suffering from a mental illness or sick-house syndrome. However, it may be impossible to require this discernment of all doctors. It became very difficult even for me to examine patients with chemical sensitivity, because today it is not the situation that the concentration of a given chemical substance is high. The patient may have been exposed to several kinds of chemical substances, and symptoms may appear. It is necessary to carry out a confirmed diagnosis derived from objective testing.

Y. Yanagisawa: I ask Dr. Miyata. What kind of symptoms do the latest patients show?

M. Miyata: Most patients' symptoms are tiredness and headache. One patient with a clear environmental condition showed symptoms of sick-house syndrome. However, although this patient developed symptoms as a child, the number of patients in whom an environmental condition is not clear has been increasing recently. As it cannot be specified when symptoms developed, I may be unable to judge a cause-and-effect relationship clearly. It is difficult to perform an oral consultation and a sufficiently detailed medical examination on a child. It is very difficult to investigate a patient's medical history and living environment in a short time.

Y. Yanagisawa: The typical symptoms of the sick-house syndrome are persistent fatigue, headache, and poor concentration. Such general physical complaints do not appear only in sick-house syndrome specifically. Interpretation of such nonspecific symptoms is subjective. What kind of objective diagnostic method is there for sick-house syndrome?

M. Miyata: The subjective index obtained by oral consultation is the most important in diagnosis of sick-house syndrome. Information regarding the point at which the condition gets worse is especially important. Features of the house or building may lead to specification of the chemical substance that worsens the condition. As a general physical complaint is not a symptom that appears only in sick-house syndrome specifically, one needs to identify whether the condition is sick-house syndrome. Therefore, it is necessary to use objective examination data. I discussed the details of the objective inspection method in Chapter 5 of this book.

Y. Yanagisawa: When the foregoing are summarized, large numbers of houses in which the concentration of formaldehyde, toluene, or *p*-dichlorobenzene is high were constructed in Japan in the 1990s. Sick-house syndrome and chemical sensitivity occurred frequently in such houses. The Ministry of Health, Labor and Welfare and the Ministry of Land, Infrastructure and Transport decided on countermeasures.

The Ministry of Health, Labor and Welfare defined the guideline value of formaldehyde first in 1997, and the guideline value of a total of 13 substances by 2002. The Ministry of Land, Infrastructure and Transport revised the Building Standards Act in 2005. The revised Building Standards Act restricted use of building materials that diffuse formaldehyde and forbade use of chlorpyrifos as a termite controlling agent.

16.3 GOVERNMENT ACTIONS

S. Ishikawa: Dr. Yanagisawa has agreed to participate in the committee that examines a guideline value, and decided the guideline value of formaldehyde as the first guideline value. Next, the guideline value of hydrocarbon systems, such as toluene and styrene, was established. The concentrations in indoor air began to decrease.

However, the number of patients did not decline at all by setting up a guideline value. I carried out an examination of the chemical substances of a hydrocarbon system, and then of an organophosphorus pesticide. I examined chlorpyrifos (CP). Many CPs are used even in the United States. It is said that in the United States one-third of people who react to a chemical substance sensitively should suspect organophosphorus agricultural chemicals. Although there was also a dissenting opinion, I claimed that use of CP should stop absolutely.

H. Yoshino: I think that revision of the Building Standards Act is greatly reflecting a change of a patient's pathological condition. When a nationwide investigation was conducted paying attention to formaldehyde, it turned out that 27.3% of residences exceeded the guideline value. From these results, to reduce the number of victims of sick-house syndrome, the Building Standards Act was revised in 2002. The chemical compounds that the

Building Standards Act designated as the control subject are formaldehyde and chlorpyrifos. Formaldehyde is used for adhesives, such as in plywood, and chlorpyrifos is used for the extermination of termites. Chlorpyrifos use is prohibited in the revision of the Building Standard Act. As for formaldehyde, the amount in the building materials used was restricted according to the amount of diffusion.

In the revised Building Standards Act, a reservation of 0.5 times or more of an air change rate and installation of machine ventilation equipment was mandated. However, if it actually measures 0.5 times, the amount of ventilation will not necessarily be filled. Although installation of a ventilator is not obligatory, the air change rate of makeshift houses constructed after a great earthquake in East Japan is insufficient. Even if a ventilator is installed, the resident may not use it in many cases. Unless a resident is conscientious, sufficient ventilation may not be performed even if ventilation equipment is available. Moreover, even if the ventilator is operated, dust gradually accumulates in the filter and ventilation becomes insufficient. Maintenance of a ventilator is very important.

By revision of the Building Standards Act, the concentration of 13 substances specified as guideline values, such as formaldehyde, decreased. However, substances that were not specified came to be used as alternative materials. When indoor TVOC concentration is measured, there are actually many cases in which very high concentrations are shown. TVOC concentration can be reduced if the amount of ventilation in a house is increased.

The failing points of the measure against sick-house syndrome in Japan are that the TVOC concentration in alternative materials is high, and that reservation of an air change rate is not fully performed. This information should prove helpful when taking measures against sick houses in readers' countries from now on.

16.4 PATIENTS' COMPENSATION

M. Miyata: It is very difficult for person with sick-house syndrome to specify a causative agent and to win in a trial. For example, a patient could be a teacher in a school where ventilation is insufficient. Such a school is called a sick school. There are many workplaces where ventilation is insufficient in winter. When symptoms develop in such an office, it may be difficult to prove a causative agent.

Y. Yanagisawa: When people develop sick-house syndrome or a chemical sensitivity, a patient asks a court for relief. However, in the logic of a court, a plaintiff needs to clarify a causative agent. That is, it is necessary to prove a scientific cause-and-effect relationship between a causative agent and a symptom. There is difficulty in clarifying a scientific cause-and-effect relationship in the present indoor air environment conditions.

Because the TVOC concentration, which is the total amount of a volatile organic compound, is high, it is difficult to clarify the cause-and-effect relationship between a given chemical compound and a symptom.

S. Ishikawa: In organophosphorus pesticides, the guideline value was decided for diazinon and dichlorvos (DDVP) as representatives. However, a guideline value was not provided for the other organophosphorus pesticides. The guideline value of a neonicotinoid insecticide or a thiocarbamate herbicide was not examined.

The guideline value of only 13 substances was defined for 5 years from 1997 to 2002. A chemical substance for which the guideline value was determined is no longer used because a construction material with the same function was developed using an alternative chemical compound. For example, agricultural chemicals, such as a neonicotinoid insecticide, were developed to replace organophosphorus pesticides. Solvents from aromatic series systems, such as toluene, are no longer used. Many aliphatic solvents came to be used instead.

Y. Yanagisawa: As a result, although TVOCs did not decrease, the concentrations of 13 substances in which the guideline value was defined fell. The situation thereby changed.

What are the most effective countermeasures? The pollution situation of indoor air and the sick house patient's symptoms became complicated because administrative measures, such as setting guideline values, progressed.

H. Yoshino: Since the alternative chemical compound was used and the pollution situation became complicated, it is important to maintain a 0.5 air change rate specified in the Building Standards Act revision.

Y. Yanagisawa: In addition to the 13 substances for which the guideline value was defined, the Ministry of Health, Labor and Welfare has defined the provisional desired value of TVOC as the 14th. TVOC is the total concentration of VOCs. Therefore, the TVOC value was not necessarily defined by the assessment of risk to health like a guideline value substance.

From the logic of a scientific cause-and-effect relationship, it has not been proved that TVOC concentration caused sick-house syndrome. Even if a plaintiff patient claims that the cause of sick-house syndrome was a high TVOC concentration, the plaintiff cannot win in a trial. In court, the plaintiff must prove which substance in TVOC caused the sick-house syndrome.

TVOC is not a useful index if such logic is followed. There is an example of a case settled in a trial in which the logic applied in reaching a verdict differed from a scientific cause-and-effect relationship. The court ruled that it was a safety manager's breach of duty that led to the development of sick-house syndrome in an employee working in an office where the TVOC concentration was high. Because it was a violation of a safety duty, the safety manager had to compensate damages.

H. Yoshino: The value 400 $\mu g/m^3$, which is a provisional desired value of TVOCs, is such a low value for construction business that realization is difficult.

Y. Yanagisawa: This trial concerned a situation in which the person who was working in the temporary office immediately after construction was exposed to a high TVOC concentration and developed sick-house syndrome. This person's desk was on the second floor of the temporary office. All the polluted air on the first floor rose to the second floor. The person who was working in the area to which the polluted air of the first floor permeated had shown symptoms of sick-house syndrome. As a result, I regard the judgment on TVOC as a violation of a safety duty as one new standard.

Japan also has damage compensation through workers' compensation. However, these days, a person who develops sick-house syndrome can hardly receive workers' compensation.

S. Ishikawa: Possibly the governmental policy changed to less supportive of SHS patients by worker's compensation system.

M. Miyata: The patient has to measure a chemical contaminant at his or her own expense. However, because measurement is started after a patient has become affected by sick-house syndrome, a long time has passed. Therefore, if a duty to measure TVOC is imposed immediately after the end of construction or modification, the cause of sick-house syndrome will be easy to prove.

Y. Yanagisawa: Well, I had an experience in which TVOC of new building or a modification residence was measured continuously. If 3 to 6 months pass after construction is completed, usually the TVOC level becomes 400 $\mu g/m^3$ or less.

Before a patient recognizes his or her condition as sick-house syndrome or chemical sensitivity, it will take from 3 to 6 months. Although the symptoms of sick-house syndrome are manifested when TVOC concentration is high, TVOC concentration is low at the time when it is recognized. When a doctor diagnoses a patient with sick-house syndrome, the TVOC concentration has fallen.

H. Yoshino: Is TVOC measured easily?

Y. Yanagisawa: The standard method has not been decided yet. Since T expresses the total, how to decide the components of TVOC is difficult. A standard method of measuring TVOC should first be developed. Then TVOC is measured immediately after new building or repair, and recording it is desirable. I think that a fair trial will take place if there is a record. At present, immediately after construction, measurement of the concentration of formaldehyde, toluene, xylene, ethyl benzene, and styrene is mandated by the law ensuring house quality. However, TVOCs are not included.

Although air pollution by a chemical compound has been the subject so far, I include the humidity of a residence as well.

16.5 DAMP BUILDINGS

H. Yoshino: Although the definition of the word "damp building" is rather difficult, an easy definition is a humid dank building. This problem attracted attention in the early 2000s in Europe. After countermeasures against the formaldehyde pollution problem were implemented and the VOC concentration provided in the guideline value decreased, the problem of mold growing in a dank building occurred. A correlation between children's allergic disease and mold was suspected and many investigations were conducted.

The word "damp building" came to be used even by the World Health Organization. Attention to it was paid even in Japan and many buildings were investigated with regard to the relation between mold and allergy. In a building in which mold developed or that had spots of water, it was observed that many children suffered from allergic diseases such as asthma.

This relation was statistically significant.

Y. Yanagisawa: Mold grows in residences that are humid and where dew condensation occurs, and an allergic subject is reported. Are there any interactions between the sick-house syndrome and the symptoms of the damp building patient?

S. Ishikawa: I'll relate a personal experience. I went to the apartment of a scholar in a prominent nerve research center in a large U.S. city. The building was what is called a luxury apartment with thick walls. I was invited to look at the back side of a dank place. I saw a large amount of mold that became a thick layer. There are apartments in which mold is growing with frightful power.

The weather of New York City is dry compared to Tokyo. Therefore, the building in New York City was not subject to a summer of high humidity like Japan. However, dampness was produced in the building and it was assumed that not much mold grows there. Northern European people are also becoming nervous about damp buildings. I think that considering the high humidity of Japan in a rainy season it is in a more severe damping condition than the West. How does Dr. Miyata consider the latest situation of mold in Japan?

M. Miyata: A long time ago, mold grew in the whole surface of a tatami during the rainy season of Japan. Unless the tatami was wiped two or three times, people were not able to inhabit the room. It seems, however, there were not many allergic persons in those days.

I think that it is rare rather to become allergic solely because of mold. Of course, damping is not good. Florence Nightingale stressed the importance of ventilation. Research is still lacking on the relation between a chemical sensitivity and an allergy attributable to mold. However, allergies may also get worse by exposure to a chemical compound.

Y. Yanagisawa: Only high humidity did not cause multiplication of mold. It is also the reason use of adhesives containing formaldehyde was stopped. There are sterilizing properties in formaldehyde and multiplication of mold is prevented.

Before use of adhesives containing formaldehyde, wallpaper was stretched with adhesives containing starch. When the wallpaper stretched by using starch glue was removed, mold grew and the reverse side of the wallpaper was a deep black. Starch was food for the mold.

However, multiplication of mold will not be observed if adhesives containing formaldehyde are used. It is because formaldehyde is diffused from the urea formaldehyde adhesives. If the adhesives containing formaldehyde absorb water, hydrolysis will occur. Formaldehyde gas is emitted by hydrolysis. The diffused formaldehyde first suppresses multiplication of mold and then pollutes indoor air.

16.6 TRADE-OFFS

S. Ishikawa: However, while mold and allergy are suppressed, a chemical sensitivity occurs with formaldehyde. It is antimony.

Y. Yanagisawa: The method of suppressing multiplication of mold without using a chemical compound is making it dry. Mold propagates where dew condensation is produced in a damp building. Dew condensation occurs when there is a difference in temperature and high humidity. To prevent it, air circulation in a house it useful so as not to create a closed space where mold can grow.

For example, if the door of a closet or furniture is opened a little, air will circulate and it will become difficult for dew condensation to occur and there is no longer an odor of mold. What indoor humidity is desirable?

H. Yoshino: According to the Building Standards Act or Act on Maintenance of Sanitation in Buildings, acceptable humidity is said to be 40% to 70%. However, there is no direction for a lower limit in a U.S. society standard (American Society of Heating, Refrigerating and Air Conditioning Engineers [ASHRAE]). I think greater than 70% humidity is considerably high.

Y. Yanagisawa: I think the same way. Humidity of 70% is too high.

H. Yoshino: In Hokkaido, indoor humidity of 30% or 20%, in winter is often observed. There is also research suggesting that it is more desirable to make the lower limit of humidity specified by law 40% or less. On the other hand, a medical doctor says that 40% or less is not good for preventing the common cold and influenza.

S. Ishikawa: There is a saying, "Don't dry the interior of a room too much." When heating in winter, about 40% is desirable, and about 50% humidity is desirable in spring and autumn. In some cases, it is said that 60% is suitable for skin.

Y. Yanagisawa: Although it is said that a higher humidity is effective in prevention of influenza, what is your opinion? In northern Europe, there are many houses at about 5% humidity in the winter. However, influenza does not spread so much in such a house.

Is high humidity effective in prevention of influenza?
The high humidity is more comfortable for a patient suffering from influenza.

M. Miyata: It is admitted that higher humidity is desirable to prevention of influenza. The population density is different between northern Europe and Japan. In densely populated Japan, many people come in close contact with influenza patients through commuting and so forth. The situation is different in northern Europe.

Y. Yanagisawa: If a residence is humid, mold will grow, but influenza is prevented. What is your opinion about the antifungal agent that prevents multiplication of mold?

M. Miyata: There is a case of patient acquiring a chemical sensitivity after using the antifungal agent. When you use a chemical compound indoors, please recognize that there is a risk of getting a chemical sensitivity.

Y. Yanagisawa: Because an antifungal agent is scattered by a spray, a user inhales mist through the lungs. The chemical ingredient in the antifungal agent passes into blood from the lungs. The chemical ingredient of an antifungal agent enters the body without undergoing detoxification. On the other hand, a substance taken in through the mouth is detoxified by the liver. The influence on health is very intense when mist is inhaled. When scattering a chemical compound by a spray, it is necessary to call attention to possible adverse health effects.

M. Miyata: I fully agree with your comment. When inhaling mist compared with volatilized compounds, the amount of absorption increases greatly. The spray agents such as an insecticide or a waterproofing spray are also very dangerous.

Y. Yanagisawa: The need to use waterproofing spray outdoors was shown clearly. A person died when using the waterproofing spray indoors. Ventilation is important so as not to inhale the mist of a chemical compound. Although a chemical compound is convenient, it is important to understand well that there are also side effects and to use it carefully.

Y. Yanagisawa: We have the freedom to choose a comfortable life style by ourselves. To have a comfortable life, we need to understand more about the side effects of our actions. Too much use of chemical agents induces more undesirable side effects. A proverb says that the last drop makes the cup run over; it is too much of a good thing. Less is more.

Dr. Ishikawa, Dr. Miyata, and Dr. Yoshino spoke from a viewpoint of specialists about sick-house syndrome, chemical sensitivity, and trade-offs.

We hope this book will serve as a source of useful information to help people lead better and healthier lives. Thank you.

Epilogue

There are two main causes of sick-house syndrome and multiple chemical sensitivity. First, various chemicals have come to be used in building materials, facilities and equipment, household goods, fragrances, insect repellents, agrichemicals, and so on. Second, the airtight performance of buildings has improved and natural ventilation is insufficient. In addition, the impact of chemicals on people varies greatly depending on genetic factors, constitution, dietary habits, environmental conditions in past residences and other areas, and so on. Some people do not experience any ill effects and some become sick after exposure to the same concentration of the chemical. To deal effectively with the disease, it is necessary to eliminate the causative chemicals as much as possible and to ventilate actively, as well as to seek medical treatment and develop resistance through improved dietary habits and moderate exercise.

The main purpose of this book is to describe methods of preventing and treating sick-building syndrome and chemical sensitivity. The book summarizes investigations and research spanning more than 10 years such as history of the disease, patients' living conditions, diagnosis and treatment of the disease, evaluation and indication of chemical emission from building materials, measurement of the concentration of chemical substances, current indoor environment in sick buildings, and architectural preventive methods. It can be said that the book presents research achievement tackled comprehensively by experts in medical science, pharmacy, chemistry, building science, and other fields.

To deal with these health problems, the Building Standards Law to prevent sick-house syndrome in Japan was revised and enforced in 2003; it limits use of building materials containing formaldehyde and requires installation of ventilation equipment. Although the number of patients consequently declined, there are still people who suffer from these diseases. In addition, with respect to the 13 chemicals, including formaldehyde, for which the Ministry of Health, Labour and Welfare issued guidelines, the indoor concentration in new housing in recent years is lower but there are many cases where the total volatile organic compound concentration is greater than 400 $\mu g/m^3$, and some people become ill. Also, the number of chemicals is 70 million or more and is still growing. There is a possibility that new chemical substances that cause health problems will emerge in the future. Therefore, it is expected that diagnostic methods, treatment, concentration measurement method, preventive measures, and so on indicated in the book will not only provide an answer to various problems associated with the current sick-house syndrome and chemical sensitivity, but also shall serve as a prescription for unknown disease causes.

Hiroshi Yoshino
Dr. of Engineering, Professor Emeritus and
President-Appointed Extraordinary Professor, Tohoku University

Index

Page numbers followed by f and t indicate figures and tables, respectively.

A

Abdollahi, M., 8
Abdominal breathing, 42–43
Accommodation examination, of eyes, 33
Accurate measurement method, *see* Integration measurement
Acetaldehyde, 110, 112, 114, 121, 121f, 123, 124, 128, 143, 146, 152, 153, 156
Acetylacetone absorptiometric method, 132
Acetylcholinesterase (AchE)
  activity, of red cells, 45
  inhibition, 5
Acetylcysteine, 40
Active charcoal system, air purifier, 38
Active method, measuring indoor concentration, 128
Active sampling (AS), 135–136, 136f
Act on Maintenance of Sanitation in Buildings, 85–87, 85f, 86f, 86t, 93, 94, 140, 141t
Acupuncture, 43
Acute toxicology, 21, 22f
Adaptation syndrome, GAS, 57–58, 58f
Additives, food, 12, 19, 37
Adhesives, 107, 109–110, 109f, 111–112, 111f, 145, 145f
Adjacent concealed spaces, restrictions concerning, 84
ADPAC, 71
Advanced diffusive sampling emission cell (ADSEC), 72–73, 135, 135f
Advanced Industrial Science and Technology (AIST), 122
Agents, chemical, 5–8, 7t
Agricultural chemicals, defined, 13
Air
  chemical substances, avoidance of, 38–39, 39f
  cleaners, 148
  fresheners, 110–111
  indoor, *see* Indoor air
  outdoor, contaminants in, 112–113
Air change rate
  formaldehyde concentration and, 104, 104f, 155, 155f
  measurement, 153
Air pollution, 97–105
  Building Standard Law (2003), amendment, 87, 87f, 88f
  chemical substance concentrations *vs.*, 104–105, 104f, 105f
  measurement results
    of houses, 100–102, 100f, 101f
    ventilation rate, 102–103, 102f
  measurement survey, 97–100
    period and investigated houses, 97, 98f, 98t
    points and method, 97, 98t, 99–100, 99f
Air purifiers, 38–39, 39f
Air-tightness, 97–105
  chemical substance concentrations *vs.*, 103–104, 103f, 104f
  formaldehyde concentration and, 153, 155f
  measurement, 153
  measurement results
    of houses, 100–102, 100f, 101f
    ventilation rate, 102–103, 102f
  measurement survey, 97–100
    period and investigated houses, 97, 98f, 98t
    points and method, 97, 98t, 99–100, 99f
Alcohols, 111
Aldehydes, 111
Aliphatic compounds, 143, 145
Allegology & Immunology, 2
Allergies
  children with, *see* Children
  conjunctivitis, 22, 22f
  drug, 39
  food, 12, 19, 62, 64, 64f
  hypersensitive reaction, 22
  to metals, 18
  symptoms, 12
  trihalomethanes and, 37
Allgemeine Ortskrankenkasse (AOK), 60
Alpha-lipoic acid, 41
Alternative medicine, for CS, 43
Alternatives, to guideline substances, 143, 145
  aliphatic compounds, increased use of, 145
  formaldehyde emission adhesives, low, 145, 145f
  water-based paint and ink, 145
Alternative therapy, for CS, 42–43
  alternative medicine, 43
  eastern medicine, 43
  prayer and meditation in Zen Buddhism, 42–43

Aluminum, 60
Alzheimer's disease, 13, 60
American Chemical Society, 53
Anemia, 60
Anesthesia, sensitive reaction, 18
Animal experiments, using extremely low dosages, 21, 22–23, 22f, 23f
Antifungal agents, 111
α-pinene, 147, 148
Appliances, burning, 112
Arsenic, 60
Artificial wooden boards, 107, 109–110, 109f
Ashford, N. A., 1
Aspirin, 18
Assessment, risk, *see* Risk assessment
Asthma, 37, 45, 46, 114, 172
Atropine, 6, 42
Attention deficit hyperactivity disorder (ADHD), 60
Attic, restrictions concerning, 84
Austrian Medical Association, 14
Autoimmune diseases, gluten sensitivity and, 64
Automatic nervous system dysfunction, 31

B

Benzene, 17, 112, 114
Benzyl butyl phthalate (BBP), 114
Biochemical individuality, 54–57, 55t–56t
Bisphenol A (BPA), 23, 62, 63t
Bromine compounds, 112
Building Act, 123, 123t
Building age, chemical substance concentration and, 159, 159f, 160f
Building and statutory regulations, ventilation strategies, 79–95
  indoor air pollution measures in Japan, 80–82
    design assumption and strategy, 81–82, 82f
    against sick houses, 80–81
  indoor air quality, statutory regulations for, 83–88
    Act on Maintenance of Sanitation in Buildings, 85–87, 85f, 86f, 86t
    attic and adjacent concealed spaces, restrictions, 84
    Building Standard Law (2003), amendment, 83, 87, 87f, 88f
    Housing Quality Assurance Act, 84
    interior finishing, restrictions on, 83
    materials, formaldehyde and installation of ventilation, 83–84
    ventilation, mandatory installation, 83–84
  mechanical ventilation systems
    with heat exchanger, 90–91
    types of, 88, 88t, 89–90
    utilizing low-power input motors, 90
  overview, 79
  technology development and energy-saving performance, 90–91
    demand-controlled ventilation, 90
    hybrid ventilation systems, 91
    mechanical ventilation systems, *see* Mechanical ventilation system
  ventilation, types of, 88–90, 88t, 89t
    hybrid, 89, 89t
    mechanical, 88, 88t, 89–90
    natural, 88, 89
  ventilation equipment in large buildings, 93–95
    centrally controlled HVAC systems, 94–95
    in designated buildings, 93–94, 94t
  ventilation systems, planning and operation, 91–93
    concerning regular maintenance, notes, 92, 93
    construction, notes on, 92
    path, 91–92
    pressure drops, notes for calculating, 92
    steps for, 93, 93f
Building performance, field survey, 152–153
Building products and materials, 69, 70, 71, 76
  chemical compounds, 70; *see also* Chemical compounds
  emission rate, methods of measuring, 131–135
    ADSEC, 135, 135f
    chamber method, 132–133, 132f
    desiccator method, 131–132, 132f
    FLEC, 133, 133f
    on-site measurement methods, 133–135
    PECS, 133, 134, 134f
    PFS, 134, 134f
  formaldehyde and installation of ventilation, 83–84
    attic and adjacent concealed spaces, restrictions, 84
    interior finishing, restrictions on, 83
    ventilation, mandatory installation, 83–84
Building Standard Law
  amendment
    in 2003, indoor air pollution, 87, 87f, 88f
    technical standards, 83
  formaldehyde regulation, designing, 81
  indoor air pollution, 80
  indoor pollution concentration, 81–82
  in Japan, 3, 79, 90
  regulatory measures of, 79
  revisions of, 85
  SBS/SHS and, 3, 75, 76
  ventilation systems, 94
Building Standards Act, revised, 140, 141, 141t, 142, 146, 168–169, 170, 173
Burning appliances, 112

Burnout syndrome, 58f, 59
Butyl acetate, 110

C

Calcium, metabolism, 12, 14
Camphor, 111
Cancer
  chemical exposure and, 53
  diagnosis, 53
  electromagnetic waves, effects, 13
  endocrine disruptors and, 61
  indoor air pollutants, 114
  inducing effect of arsenic, 60
Carbonyl compounds
  integration measurement, 128–129, 129f
  measurement, 97, 152–153
Catalyst system, air purifier, 38
Causative chemicals, determination of, 34, 35–36, 35t
Cement, hyperreaction to, 18–19
Centrally controlled HVAC systems, 94–95
Cerebral function examination, 33, 33f
Challenge test, 34, 35–36, 35t
Chamber, emission tests using, 70–75
  desiccator method, 72, 72f
  large-chamber method, 71, 72f
  microchamber method for SVOCs, 73–75, 74f
  passive method, 72–73, 73f
  small-chamber method, 70–71, 71f, 76
Chamber method, 69, 70, 132–133, 132f
Chemical Abstracts Service (CAS), 53
Chemical compounds, emission rate, 69–76
  building products and materials, 70
  categories, 69–70, 70t
  emission tests using chamber, 70–75
    desiccator method, 72, 72f
    large-chamber method, 71, 72f
    microchamber method for SVOCs, 73–75, 74f
    passive method, 72–73, 73f
    small-chamber method, 70–71, 71f
  labeling, 75–77
    formaldehyde, 75, 76t
    VOCs, 76, 76t
  overview, 69
Chemical compounds, usage and sources, 107–113
  artificial, 108f
  artificial wooden boards and adhesives, 107, 109–110, 109f
  contaminants in outdoor air, 112–113
  fungicides, repellents, mothballs, air fresheners, and disinfectants, 110–111
  indoor source, contribution, 108f
  paint, 110, 110f
  plasticizers and flame retardants, 111–112, 111f
  secondary generated compounds, 112
  smoking and burning appliances, 112
  unknown pollutants, 113
Chemical load tests, 34–36
  causative chemicals (challenge test), 34, 35–36, 35t
  clean room, use of, 34, 35f
Chemical sensitivity (CS)
  agents in, 5–8, 7t
  background, 1–3
  causative compounds, search for, 135–136, 136f
  in children, *see* Children
  clinical symptoms, 3–4, 4t
  diagnosis, *see* Diagnosis
  disease registration, 2–3
  in Europe, 3
  in Japan, *see* Japan
  MCS, medical aspects from Germany, *see* Multiple chemical sensitivity (MCS)
  medical facts, *see* Medical facts
  on patients' daily lives, *see* Patients, daily lives
  psychiatric and social aspects, 4–5
  SBS/SHS, *see* Sick-building/house syndrome (SBS/SHS)
  treatment, *see* Treatment, of CS
  understanding
    lack of, 8
    promoting, *see* Promoting, understanding of CS
Chemical substance(s)
  avoidance of, 37–39
    air, 38–39, 39f
    air purifiers, 38–39, 39f
    clothes, 39
    food, 37
    medicine, 39
    water, 37–38
  categories, 69–70, 70t
  concentrations
    air change rate *vs.*, 104–105, 104f, 105f
    air-tightness *vs.*, 103–104, 103f, 104f
    building age and, 159, 159f, 160f
    countermeasures against mitigation of SHS, 161–162, 161f
    countermeasures against SHS, 162–163, 162f, 163f
    long-term observations, 161–163
    measurements of, 152–153, 154t
    severity of symptoms and, 157, 158t
    SHS and, relationship, 156, 157f
Children, CS in, 45–51
  formaldehyde concentration and symptoms, 45

general symptoms, 45–46
indoor chemicals on intelligence and cognitive function, 47–50
intelligence, evaluation of, 47, 48f, 49f
SBS/SHS, development, 49–50, 50f
WISC-III test results and concentration, 47, 48, 49f
RChE, 46
sumatriptan succinate and SSRIs, case report, 50–51, 51f
TVOC concentration and symptoms, 46
Chinese medicine, 43
Chinese restaurant syndrome, 43
Chlordane, 111
Chlorination, of water, 37
Chlorpyrifos, 8, 69, 79, 83, 111, 123, 129, 141, 152, 168–169
Cholinesterase (ChE), 6
Chronic intoxication, understanding, 18, 21
Cleaning, regular maintenance, 92, 93
Clean room, use of, 34, 35f
Clinical symptoms, CS, 3–4, 4t
Clothes, avoidance of chemical substances, 39
Coenzyme Q_{10}, 41
Cognitive function, indoor chemicals on, 47–50
Colborn, Theo, 114
Common ventilation fans, 89
Compensation, patients, 169–171
Comprehensive Organix Test, 60
Concentration(s)
chemical substance
air change rate *vs.*, 104–105, 104f, 105f
air-tightness *vs.*, 103–104, 103f, 104f
building age and, 159, 159f, 160f
countermeasures against mitigation of SHS, 161–162, 161f
countermeasures against SHS, 162–163, 162f, 163f
long-term observations, 161–163
measurements of, 152–153, 154t
severity of symptoms and, relationship, 157, 158t
SHS and, relationship, 156, 157f
symptoms and, 156–157
formaldehyde
air change rate and, relationship, 104, 104f, 155, 155f
air-tightness measurement and, relationship, 103, 103f, 153, 155f
building age and, 159
in New Residential Housing, 142, 143f
sick-house problem, 165, 166
symptoms and, 45
of guideline substances, 143, 144t
indoor, methods for measurement, 127–131
instant measurement, *see* Instant measurement
integration measurement, *see* Integration measurement
overview, 127
OP pesticide, 113
p-dichlorobenzene, 47, 48
TVOCs, 103–104, 104f, 105, 105f, 146–147, 165–166, 170–171
building age and, 159, 159f, 160f
regulation of, 148–149
symptoms and, 46
temporal change of indoor, 146
WISC-III test results and, 47, 48, 49f
Conjugation phase, of detoxification, 40
Conjunctivitis, 22, 22f
Construction, notes on, 92
Contaminants, in outdoor air, 112–113
Contrast sensitivity examination, of higher visual centers, 33
Counseling, for patients and new occupants, 150
Countermeasures
against mitigation of SHS, 161–162, 161f
patients, daily lives, 12–14
daily products, caution in use, 12–13
electromagnetic waves, effects of, 13–14
environmental chemicals, avoidance of, 13
food and digestive issues, 12
home, workplace, and school environment, 12
against SHS, 162–163, 162f, 163f
Countermeasures, for indoor air pollution
required/recommended, 148–149
elimination period, enforcement of, 149
information from medical doctors, collecting, 149
information sharing, 149
new guideline substances, selection method, 149
TVOC concentration, regulation of, 148–149
routine, 147–148
air cleaners, 148
oxidative decomposition, 148
secondary pollutants, 148
CS, *see* Chemical sensitivity (CS)
Cullen, M. R., 1
Cytochrome P (CYP), 24

D

Daily products, caution in use, 12–13
Damp buildings, 172–173
Decabromodiphenylether, 112
Defective smooth pursuit eye movement, 30, 30f, 31f
Demand-controlled ventilation, 90
Dentists, education for, 18–19

Desiccator, 69, 70, 72, 75, 76, 77
Desiccator method, 72, 72f, 131–132, 132f
Design assumption and strategy, of measures against indoor air pollution problem, 81–82, 82f
Designated buildings, ventilation in, 93–94, 94t
Desorption test, 75
Detoxification
CS mechanism, 23–25, 24f
HEPAR-TOX, 65–66
treatment, CS, 39–42
conjugation phase, 40
drug treatment, 40–42
EDTA, penicillamine, PAM, and others, 42
minerals, 41–42, 42t
spa, hot spring, bath, low-temperature sauna, 40
symptomatic therapy, 41
vitamins, 40–41, 41f
Diagnosis, of CS, 5, 8, 29–36
chemical load tests, 34–36
causative chemicals (challenge test), 34, 35–36, 35t
clean room, use of, 34, 35f
detailed and careful interview, 29, 30t
neuro-ophthalmological and neurological examinations, 29, 30–34
accommodation examination of eyes, 33
cerebral function examination, 33, 33f
contrast sensitivity, of higher visual centers, 33
immune examination, 34
peripheral venous blood oxygen concentration, 34
pupillary function, 31–32, 31f, 32f
respiratory function, 34
standing ability, 32
tracking eye movement examination, 30, 30f, 31f
Diagnosis, of SBS/SHS, 29, 47, 167–168
Diagnostic program, for CS patients, 59–65
BPA, 62, 63t
hidden food allergies, 62, 64, 64f
gluten sensitivity, 64
metabolic and nutritional status (organic acid profiles), toxins, 59–60
phthalates and parabens profile, 61–62
stool and digestive analysis, 64–65
toxic metals, 60–61
Diazinon and dichlorvos (DDVP), guideline value for, 170
Dibutyl phthalate (DBP), 114, 129
Dichlorodiphenyltrichloroethane, 50
Diethylhexyl phthalate (DEHP), 62, 111, 112, 114, 129
Digestive analysis, 64–65
Diisobutyl phthalate (DiBP), 114
Di-isononyl phthalate (DiNP), 111
Dilatation, pupil, 31, 31f
2,3-Dimercapto-1-propanesulfonic acid (DMPS), 65
2,4-Dinitrophenyl hydrazine (DNPH), 128–129, 129f
Disease registration, of CS, 2–3
Disinfectants, 110–111
DNA concept, therapeutic strategy, 65
Doctors, education of, 18
Dose–response curve, 18, 21, 22f, 23, 23f
D-Penicillamine, 42
Drug allergies, 39
Drug treatment, CS, 40–42
conjugation phase, 40
EDTA, penicillamine, PAM, and others, 42
minerals, 41–42, 42t
symptomatic therapy, 41
vitamins, 40–41, 41f
Duct ventilation fans, 89
Dynamic headspace methods, 69, 70
Dysfunctions, of eye movement, 30, 30f, 31f

E

Eastern medicine, for CS, 43
Eco-Hotel Castle Wolfhagen, 66–67, 67f
Ecological architecture, 66–67, 67f
Education, 18–19
for dentists, 18–19
for healthcare team, 19
laypersons, 19
medical doctors, 18
patient, for comfortable daily life, 19
Electric sensitivity (ES)
CS patients and, 13–14
ICD 10 and, 14
medical treatments, 14
symptoms, 13–14
Electromagnetic field (EMF) syndrome, 14
Electromagnetic waves, effects of, 13–14
Electromyogram (EMG), abnormalities in, 8
Elimination period, enforcement of, 149
Emission rate, chemical compounds, 69, 71, 72, 73, 75, 76; *see also* Chemical compounds
chamber method, 70
formaldehyde, 75, 76t, 123, 123t
tests, using chamber, 70–75
desiccator method, 72, 72f
large-chamber method, 71, 72f
microchamber method for SVOCs, 73–75, 74f
passive method, 72–73, 73f
small-chamber method, 70–71, 71f

Emission rate of building materials, methods of measuring, 131–135
chamber method, 132–133, 132f
desiccator method, 131–132, 132f
on-site measurement methods, 133–135
ADSEC, 135, 135f
FLEC, 133, 133f
PECS, 133, 134, 134f
PFS, 134, 134f
Endocrine disruption, 61, 114
Energy-saving performance, 90–91
demand-controlled ventilation, 90
hybrid ventilation systems, 91
mechanical ventilation system
with heat exchanger, 90–91
utilizing low-power input motors, 90
Environmental chemicals, avoidance of, 13
Environmental medicine, 66–67, 67f
Environmental Protection Agency, 94
Environmental stressors, response to, 57–58, 58f
Environments, indoor, *see* Indoor environments
Equipment, ventilation
centrally controlled HVAC systems, 94–95
in designated buildings, 93–94, 94t
in large buildings, 93–95
Ethyl acetate, 110
Ethyl alcohol disinfection, 19
Ethylbenzene, 63t, 76t, 87, 110, 114, 123, 124, 146, 156
Ethylenediaminetetraacetic acid (EDTA), 42, 65
2-Ethyl 1-hexanol, 112
Europe, CS studies in, 3
Evaluation, of intelligence, 47, 48f, 49f
Examinations
formaldehyde load, 34, 35–36, 35t
neuro-ophthalmological and neurological, in Japan, 29, 30–34
accommodation, of eyes, 33
cerebral function, 33, 33f
contrast sensitivity, of higher visual centers, 33
immune, 34
peripheral venous blood oxygen concentration, 34
pupillary function, 31–32, 31f, 32f
respiratory function, 34
standing ability, 32
tracking eye movement, 30, 30f, 31f
Exposure assessment, indoor pollutants, 120–121, 121f
Eyes, examination of
accommodation, 33
contrast sensitivity, of higher visual centers, 33
movement, 30, 30f, 31f
pupillary function, 31–32, 31f, 32f
standing ability, 32

F

F**, 75, 76
F***, 75, 76
F****, 75, 76
Fatty acid amide hydrolase (FAAH), 6
Fenthion, 22
Fiberboard, 107, 109, 109f, 145
Field and laboratory emission cell (FLEC), 69, 70, 133, 133f
Field survey, indoor pollution, 114–118, 115t–117t
Field survey, SHS and, 151–157
chemical substance concentration and symptoms, 156–157
severity of symptoms and, 157, 158t
SHS and, 156, 157f
investigation, description, 151–153
air-tightness and air change rate, measurements of, 153
building performance and indoor environment, 152–153
chemical substance concentration, measurements of, 152–153
contents of survey questionnaire, 152
investigated houses, 151–152, 152t
survey questionnaire about health conditions, 153
results, 153–156
chemical substance concentration measurements, 153, 154t
formaldehyde concentration and air change rate, 155, 155f
formaldehyde concentration and air-tightness measurement, 153, 155f
QEESI® questionnaire, 156, 156f
Finger-pressure treatment, 43
Flame photometric detection (FPD), 129
Flame retardants, 111–112, 111f
FLEC (field and laboratory emission cell), 69, 70, 77, 133, 133f
Fluvoxamine maleate, 51
Folic acid, for CS, 41
Food additives, 12, 19, 37
Food allergies, 12, 19, 62, 64, 64f
Food and digestive issues, CS and, 12
Formaldehyde, 69, 70, 71, 72, 75, 76, 77
in adhesives, 111, 113
air change rate and, relationship, 104, 104f
air-tightness and, relationship, 103, 103f, 153, 155f
carcinogenic, 114
in classroom, 124
concentration
air change rate and, relationship, 104, 104f, 155, 155f
air-tightness measurement and, relationship, 153, 155f

building age and, 159
in New Residential Housing, 142, 143f
sick-house problem, 165, 166
symptoms and, 45
emission(s)
adhesives, low, 145, 145f
from building materials, 131–132
labeling, 75, 76t
rates, 123, 123t, 131–132, 133, 134
indoor guideline value for, 118–120, 119f, 120t
instant measurement, 130
labeling, advantageous effect of, 142, 143f
load examination, 34, 35–36, 35t
measurement method, 85
new building construction, 13
primary emission sources of, 109–110
regulation, 81–82, 83
attic and adjacent concealed spaces, restrictions, 84
building materials, use, 83–84
interior finishing, restrictions on, 83
ventilation, mandatory installation, 83–84
symptoms caused by, 167
Formaldehyde dehydrogenase (FALDH), 133
Formazan, 133, 134
Fukushima nuclear power plant (Japan), 5
Full-scale intelligence quotient (FIQ), 47
Fungicides, 110–111

G

Gas chromatography (GC), 71, 128, 129
Gas chromatography (GC)/Mass spectrometry (MS), 71, 73
Gastrointestinal function profile, 55t–56t, 57
GC-FID, 71
Gene polymorphism, of GST, 24
General adaptation syndrome (GAS), 57–58, 58f
Germany, medical aspects from
MCS, *see* Multiple chemical sensitivity (MCS)
Gibson, P. R., 4
Glia cells, 54
Gluculonosyl transferase, 24
Glucuronate, 40
Glucuronic acid, 40
Glutathione, 6, 40
Glutathione-*S*-transferase (GST), 24, 24f
Gluten sensitivity, 64
Government actions, 168–169
Grading, formaldehyde emissions, 75
Grimm, Jacob, 66–67
Grimm, Ludwig Emil, 66–67
Grimm, Wilhelm, 66–67
Guideline substances, concentrations of, 143, 144t
Guideline values
for DDVP, 170
indoor density, 141, 142t
in OP pesticides, 170
trends after determination, 142–148
aliphatic compounds, 145
alternative substances, 143, 145, 145f
formaldehyde labeling, advantageous effect of, 142, 143f
high concentration substances, 143
low formaldehyde emission adhesives, 145, 145f
social recognitions, 146–148; *see also* Social recognitions
substances, concentrations of, 143, 144t
TVOC concentration, temporal change of indoor, 146
VOCs, composition of, 143
water-based paint and ink, 145
Guinea pig experiments, 37
Gulf War syndrome, 24

H

Hazard assessment, of chemical compounds, 118–120, 119f, 120t
Hazard quotient (HQ), 121–122
Healthcare team, education for, 19
Health effects, determination of pollutants, 113–118
field survey, 114–118, 115t–117t
target, 113–114
asthma, 114
cancer, 114
endocrine disruption, 114
SBS/SHS, 113–114
Health insurance, in Japan, 17
Heart rate variability (HRV), 136
Heat exchanger, mechanical ventilation system with, 90–91
Heating, ventilating, and air conditioning (HVAC) systems
centrally controlled, 94–95
ventilation fans in, 90
Heavy metals, toxicity, 60–61
HEPAR-TOX detoxification, 65–66
Hidden food allergies, 62, 64, 64f
Higher visual centers, contrast sensitivity examination of, 33
High-pressure liquid chromatograph (HPLC), 71, 128
Hot bathing, 40
Hot spring, 40
Housing Quality Assurance Act, 79, 84, 123, 139, 140t, 146

Hubbard Regimen, 40
Hybrid ventilation systems, 89, 89t, 91
Hyperreaction, to cement, 18–19
Hypersensitivity, of human body, 1
Hypnotics, 41

I

Imidazoles, 111
Imigran®, 50
Immune examination, 34
Indoor air pollution, current situation, 139–150, 165–166
  counseling/inquiry systems for patients and new occupants, 150
  countermeasures, required/recommended, 148–149
    elimination period, enforcement of, 149
    information from medical doctors, collecting, 149
    information sharing, 149
    new guideline substances, selection method, 149
    TVOC concentration, regulation of, 148–149
  guideline values, trends after determination, 142–148
    aliphatic compounds, increased use of, 145
    alternative substances, 143, 145, 145f
    formaldehyde emission adhesives, low, 145, 145f
    formaldehyde labeling, advantageous effect of, 142, 143f
    high concentration substances, 143
    social recognitions, 146–148; *see also* Social recognitions, of indoor air pollution
    substances, concentrations of, 143, 144t
    TVOC concentration, temporal change of indoor, 146
    VOCs, composition of, 143
    water-based paint and ink, 145
  laws and regulations, 139–142
    Act on Maintenance of Sanitation in Buildings, 140, 141t
    Housing Quality Assurance Act, 139, 140t
    indoor density guideline value, 141, 142t
    Revised Building Standards Act, 140, 141, 141t
    School Health and Safety Act, 139, 140t
Indoor air pollution, measures in Japan, 80–82
  design assumption and strategy, 81–82, 82f
  against sick houses, 80–81
Indoor air quality, statutory regulations for, 83–88
  building materials containing formaldehyde, 83–84
    attic and adjacent concealed spaces, restrictions, 84
    interior finishing, restrictions on, 83
    ventilation, mandatory installation, 83–84
  Building Standard Law (2003), amendment, 87, 87f, 88f
  technical standards
    Act on Maintenance of Sanitation in Buildings, 85–87, 85f, 86f, 86t
    amended Building Standard Law, 83
    Housing Quality Assurance Act, 84
Indoor chemicals
  on intelligence and cognitive function, 47–50
  WISC-III test results and concentration, 47, 48, 49f
Indoor density guideline value, 141, 142t
Indoor environments
  chemical species, usage and sources, 107–113
    artificial chemical compounds, 108f
    artificial wooden boards and adhesives, 107, 109–110, 109f
    contaminants in outdoor air, 112–113
    contribution of indoor source, 108f
    fungicides, repellents, mothballs, air fresheners, and disinfectants, 110–111
    paint, 110, 110f
    plasticizers and flame retardants, 111–112, 111f
    secondary generated compounds, 112
    smoking and burning appliances, 112
    unknown pollutants, 113
  in SBS/SHS, investigation, 151–163
    field survey, 151–157; *see also* Field survey
    investigation, description of, 152–153
    long-term observations, 157–163; *see also* Observations, long-term
Indoor pollutants, chemical features, 107–124
  chemical species, usage and sources, 107–113
    artificial chemical compounds, 108f
    artificial wooden boards and adhesives, 107, 109–110, 109f
    contaminants in outdoor air, 112–113
    fungicides, repellents, mothballs, air fresheners, and disinfectants, 110–111
    indoor source, contribution of, 108f
    paint, 110, 110f
    plasticizers and flame retardants, 111–112, 111f
    secondary generated compounds, 112
    smoking and burning appliances, 112
    unknown pollutants, 113

determination, health effects and, 113–118
asthma, 114
cancer, 114
endocrine disruption, 114
field survey, 114–118, 115t–117t
SBS/SHS, 113–114
target, 113–114
enforcement and effect of regulations, 123–124
impact, 124, 124f
time course, 123, 123t, 124
risk assessment, 118–122
defined, 118
exposure, 120–121, 121f
hazard, 118–120, 119f, 120t
HQ and MOE, 121–122
Indoor pollution, methods for measurement, 127–136
concentration, 127–131
instant measurement, *see* Instant measurement
integration measurement, *see* Integration measurement
overview, 127
emission rate of building materials, 131–135
ADSEC, 135, 135f
chamber method, 132–133, 132f
desiccator method, 131–132, 132f
FLEC, 133, 133f
on-site measurement methods, 133–135
PECS, 133, 134, 134f
PFS, 134, 134f
personal exposure amount, 135–136
AS–PS method, 135–136, 136f
causative compound of CS, search for, 135–136, 136f
instant measurement method, 136
TVOCs and HRV, 136
Information
from medical doctors, collecting, 149
sharing, 149
Infrared pupillometer test, 50
Ink, water-based, 145
Inquiry systems, for patients and new occupants, 150
Installation, ventilation
building materials containing formaldehyde and, 83–84
attic and adjacent concealed spaces, restrictions, 84
interior finishing, restrictions on, 83
mandatory, 83–84
Instant measurement
defined, 127
integration measurement and, 130–131
peak capture method, 131, 131f
on-site method, 130
overview, 130
personal exposure measurement method
TVOCs and HRV, 136
subject compounds, TVOCs, formaldehyde, 130
Insurance, health, 17
Integration measurement, 127–129
active method, 128
carbonyl compounds, 128–129, 129f
defined, 127
instant measurement and, combined application, 130–131
peak capture method, 131, 131f
overview, 127–128
passive method, 128
subject compounds, 128–129
SVOCs, 129
VOCs and TVOCs, 129
Intelligence
evaluation, 47, 48f, 49f
function, indoor chemicals on, 47–50
Interior finishing, restrictions on, 83
Intermediate syndrome, 21
International Association of Eye Research symposium, 6
International Classification of Disease 10 (ICD-10), ES and, 14
Intestinal dysbiosis, 57
Investigated houses
air-tightness of, 100–102, 100f, 101f
indoor environments in SHS
field survey, 151–152, 152t
long-term observations, 159
survey period and, 97, 98t
ventilation systems, types, 97, 98f
Investigation, of indoor environments and occupants' health in SBS/SHS, 151–163
field survey, 151–157
chemical substance concentration and symptoms, 156–157
description of investigation, 151–153
results, 153–156
long-term observations, 157–163
chemical substance concentrations, 161–163
investigated houses, 159
results, 159–161
Irritable bowel syndrome, 54
Ishikawa, S., 3, 6, 14, 29
Issues, food and digestive, 12

J

Japan
air-tightness of houses in, 97
Building Standard Law in, 3, 79, 90

cases, 3–4, 4t
diagnostic criteria, 29, 30t
ES patients and, 13, 14
formaldehyde emissions, labeling, 75, 76t
formaldehyde load examination, 34, 35–36, 35t
Fukushima nuclear power plant, 5
indoor air pollution measures, 80–82
design assumption and strategy, 81–82, 82f
against sick houses, 80–81
indoor concentrations of VOCs in, 114–118, 115t–117t
indoor guideline values in, 120t
JAS, 72, 132
JIS, 70–71, 71f, 72, 73, 76, 111, 132
large-chamber method, 71, 72f
laws and regulations, 139–142
Act on Maintenance of Sanitation in Buildings, 140, 141t
Housing Quality Assurance Act, 139, 140t
indoor density guideline value, 141, 142t
Revised Building Standards Act, 140, 141, 141t
School Health and Safety Act, 139, 140t
medical correlations, 167–168
medical insurance in, 17
neuro-ophthalmological and neurological examinations, 29–30–34
accommodation examination of eyes, 33
cerebral function examination, 33, 33f
contrast sensitivity, of higher visual centers, 33
immune examination, 34
peripheral venous blood oxygen concentration, 34
pupillary function, 31–32, 31f, 32f
respiratory function, 34
standing ability, 32
tracking eye movement examination, 30, 30f, 31f
SBS/SHS, 3, 8, 69
small-chamber method, 70–71, 71f
studies, 3
Japanese Agricultural Standards (JAS), 72, 132
Japanese Industrial Standard (JIS), 70–71, 71f, 72, 73, 76, 111, 132
JIS A 1460, 72, 77
JIS A 1901, 76, 77
JIS A 1903, 73, 77
Jonathan Foeman Award, 3, 6
Judicial level, understanding of CS, 17–18

K

Kilburn, K. H., 29
Kindling phenomenon, 26

L

Labeling, 75–77
formaldehyde, 75, 76t
advantageous effect of, 142, 143f
VOCs, 76, 76t
Laminate lumber, 109, 109f
Large buildings, ventilation equipment in, 93–95
Large-chamber method, 69, 71, 72f
Law of Arndt-Schulz, 21
Laws, in Japan, 139–142
Act on Maintenance of Sanitation in Buildings, 140, 141t
Housing Quality Assurance Act, 139, 140t
indoor density guideline value, 141, 142t
Revised Building Standards Act, 140, 141, 141t
School Health and Safety Act, 139, 140t
Laypersons, education, 19
Lead, 60
Leaky gut, defined, 64
Load examination, formaldehyde, 34, 35–36, 35t
Logan, Heather, 54
Low-dosage exposure sensitivity syndrome (LESS), 1
Low-temperature sauna, 40
Lymphocyte transformation test (LTT), on food, 62, 64, 64f

M

Magnesium, 42, 42t
Malathion®, 6
Margin of exposure (MOE), 121–122
Masking, defined, 167
Mass spectrometer (MS), 128, 129
MCS, *see* Multiple chemical sensitivity (MCS)
Measurements
chemical substance concentration, 152–153, 154t
formaldehyde concentration and air-tightness, relationship, 153, 155f
indoor air pollution, in Japan, 80–82
design assumption and strategy, 81–82, 82f
against sick houses, 80–81
of indoor pollution, methods for, *see* Indoor pollution, methods for measurement
VOCs, 97
Measurements, ventilation, air-tightness, and air pollution
results, 100–103
air-tightness of houses, 100–102, 100f, 101f
ventilation rate, 102–103, 102f

survey, 97–100
period and investigated houses, 97, 98f, 98t
points and method, 97, 98t, 99–100, 99f
Mechanical ventilation system
formaldehyde concentration and, 104, 104f
with heat exchanger, 90–91
methods of selecting, 89–90
types, 88, 88t
utilizing low-power input motors, 90
Mechanism of CS, research, 23–26
detoxification, 23–25, 24f
nervous sensitivity acquisition, 25–26, 25f
Mediators, 59
Medical correlations, in Japan, 167–168
Medical doctors
education of, 18
information from, 149
Medical facts, 21–26
acute toxicology, 21, 22f
animal experiments using extremely low dosages, 21, 22–23, 22f, 23f
chronic toxicology, 21
intermediate syndrome, 21
research on mechanism, 23–26
detoxification, 23–25, 24f
nervous sensitivity acquisition, 25–26, 25f
Medical insurance, in Japan, 17
Medicine, avoidance of chemical substances, 39
Meditation, in Zen Buddhism, 42–43
Melamine, 13, 109, 145
Mental tranquility, 42–43
Mercury, 60
Metabolic status, 59–60
Metabolism, of calcium, 12
Metals
allergy to, 18
toxicity, 60–61
Methods
chamber, 69, 70, 132–133, 132f
desiccator, 72, 72f, 131–132, 132f
dynamic headspace, 69, 70
large-chamber, 71, 72f
measurement
air pollution, 97, 98t, 99–100, 99f
air-tightness, 97, 98t, 99–100, 99f
formaldehyde, 85
indoor pollution, *see* Indoor pollution, methods for measurement
microchamber method for SVOCs, 73–75, 74f
passive, 72–73, 73f
selection
mechanical ventilation system, 89–90
new guideline substances, 149
small-chamber, 70–71, 71f, 76
Methyl cobalamin, 41
Methyl mercury, 3, 5
MHLW (Ministry of Health, Labor and Welfare), 123, 141, 146, 165, 168
Microchamber, 69, 70, 73, 74, 75
Microchamber method, for SVOCs, 73–75, 74f
Miller, C. S., 1, 29
Minerals, for CS, 41–42, 42t
Ministry of Environment, 122
Ministry of Health, Labor and Welfare (MHLW), 123, 141, 146, 165, 168
Ministry of Land, Infrastructure, Transport and Tourism (MLIT), 81, 123, 124, 139, 142
Ministry of Land, Infrastructure, Transport and Trip, 143, 147, 168
Monosodium glutamate, 43
Mothballs, 110–111
Multiple chemical sensitivity (MCS), 1, 53–67
biochemical individuality, 54–57, 55t–56t
causes, 59
CS, 54
diagnostic program, 59–65
BPA, 62, 63t
gluten sensitivity, 64
hidden food allergies, 62, 64, 64f
metabolic and nutritional status (organic acid profiles), toxins, 59–60
phthalates and parabens profile, 61–62
stool and digestive analysis, 64–65
toxic metals, 60–61
ecological architecture and environmental medicine, 66–67, 67f
environmental stressors, response to, 57–58, 58f
HEPAR-TOX detoxification, 65–66
overview, 53
Randolph's specific adaptation syndrome, 57–59, 58f
therapy, DNA concept, 65
Wolfhagen, Germany, 66–67, 67f
Multivitamins, for CS, 41

N

Naphthalenes, 111
National Health and Nutrition Examination Survey (NHANES), 62
National Institute of Technology and Evaluation (NITE), 122
National Institutes of Health (NIH), in United States, 6
Natural ventilation, 88, 89
Near-infrared spectroscopy (NIRS), 33, 33f, 51, 51f
Nervous sensitivity acquisition, 25–26, 25f
Neurological diseases, gluten sensitivity and, 64

Neuro-ophthalmological and neurological examinations, in Japan, 29, 30–34
accommodation examination of eyes, 33
cerebral function examination, 33, 33f
contrast sensitivity, of higher visual centers, 33
immune examination, 34
peripheral venous blood oxygen concentration, 34
pupillary function, 31–32, 31f, 32f
respiratory function, 34
standing ability, 32
tracking eye movement examination, 30, 30f, 31f
Neurotoxic esterase (NTE), 24
New Energy and Industrial Technology Development Organization (NEDO), 114–118, 115t–117t
New guideline substances, selection method, 149
Nicotinamide adenine dinucleotide (NAD), 133
NIRS (near-infrared spectroscopy), 33, 33f, 51, 51f
Nitrogen oxide, 38
Nonregulated substances, widespread indoor air pollution and, 146–147, 147f
No observed adverse effect level (NOAEL), 118–120, 122
Nutritional instruction, for CS, 43–44
Nutritional status, 59–60

O

Observations, long-term, 157, 159–164
chemical substance concentrations, 161–163
countermeasures against mitigation of SHS, 161–162, 161f
countermeasures against SHS, 162–163, 162f, 163f
investigated houses, 159
results of investigation, 159–161
chemical substance concentration and building age, 159, 159f, 160f
occupants' symptoms, yearly changes of, 160–161, 160f, 161f
Occupants
classifications of, 156
counseling/inquiry systems for, 150
health in SBS/SHS, 151–163
field survey, 151–157; *see also* Field survey
long-term observations, 157–163; *see also* Observations, long-term
symptoms, yearly changes of, 160–161, 160f, 161f
Ocular accommodation, 32
On-site measurement methods, 133–135
ADSEC, 135, 135f
FLEC, 133, 133f
PECS, 133, 134, 134f
PFS, 134, 134f
source intensity and ventilation rate, 130
OP compounds, *see* Organophosphorous (OP) compounds
Operation, of ventilation system, 91–93
concerning regular maintenance, notes, 92, 93
construction, notes on, 92
pressure drops, notes for calculating, 92
ventilation path, 91–92
Organic acid profiles, 59–60
Organochlorine pesticide, 50
Organophosphorous (OP) compounds
chlordane, 111
chlorpyrifos, 8, 69, 79, 83, 111, 123, 129, 141, 152, 168, 169
chronic toxicity of, 50
concentration, 113
in daily lives, 5–6
electromagnetic waves and, 14
exposure to, 46
fenthion, 22
guideline values in, 170
induced intermediate syndrome, 8
intoxication, 6, 8, 21
measurements, 152–153
NTE activity, 24
pesticides, 3, 5–6, 45
trichlorfon, 22
Osler, William, Sir, 54
Outdoor air, contaminants in, 112–113
Oxidative decomposition, 148
Oxygen concentration, peripheral venous blood, 34
Ozone oxidization system, air purifier, 38

P

Paint
chemical substances in, 110, 110f
water-based, 145
Parabens, 61–62
Particle board, 107, 109f, 145
Passive emission colorimetric sensor/sampler (PECS), 107, 133, 134, 134f
Passive flux sampler (PFS), 107, 134, 134f
Passive method, 72–73, 73f, 128
Passive sampling (PS), 135–136, 136f
Patch tests, 18
Path, ventilation, 91–92
Patients
compensation, 169–171
education, for comfortable daily life, 19

Patients, daily lives, 11–14
countermeasures, 12–14
daily products, caution in use, 12–13
electromagnetic waves, effects of, 13–14
environmental chemicals, avoidance of, 13
food and digestive issues, 12
home, workplace, and school environment, 12
effects, 11, 12f
p-Dichlorobenzene, 47, 48, 50, 111, 114, 119
Peak capture method, 131, 131f
PECS (passive emission colorimetric sensor/sampler), 107, 133, 134, 134f
Penicillamine, 42
Performance intelligence quotient (PIQ), 47, 49, 50
Period, measurement survey, 97, 98f, 98t
Peripheral venous blood oxygen concentration, 34
Peroxynitrite, 2
Personal exposure amount, methods for measurement, 135–136
AS–PS method, 135–136, 136f
causative compound of CS, search for, 135–136, 136f
instant measurement method
TVOCs and HRV, 136
Phenols, 111, 145
Phosphates, 111, 112
Photoionization detector (PID), 130
Phthalates, 61–62, 111, 114, 129
Planning, of ventilation system, 91–93
concerning regular maintenance, notes, 92, 93
construction, notes on, 92
pressure drops, notes for calculating, 92
steps for, 93, 93f
ventilation path, 91–92
Plasticizers, 61–62, 111–112, 111f
Plywood, 107, 109f, 145
Pollutants
indoor, *see* Indoor pollutants
secondary, 148
Pollution
air, *see* Air pollution
indoor, *see* Indoor pollution
air, *see* Indoor air pollution
Polyvinyl chloride (PVC), 13, 62
POM, 70
Porphyrin testing, 60–61
Pralidoximemethiodide (PAM), 6, 42
Prayer, in Zen Buddhism, 42–43
Precipitating events, defined, 59
Pressure drops, notes for calculating, 92
Products, in daily use, 12–13
Promoting, understanding of CS, 17–19
education, 18–19
for dentists, 18–19
for healthcare team, 19
laypersons, 19
medical doctors, 18
patient, for comfortable daily life, 19
at judicial level, 17–18
medical insurance in Japan, 17
Psychiatric aspects, CS, 4–5
Psychogenetic theory, CS, 2
Pulifinium bromide, 6
Pupillary reaction test, 35, 35t
Pyrethroids, 111

Q

Questionnaire
for medical examination, 1
QEESI®, 1, 29, 153, 156, 156f, 160, 160f
survey
about health conditions, 153
contents of, 152
Quick Environmental Exposure and Sensitivity Inventory (QEESI®) questionnaire, 1, 29, 153, 156, 156f, 160, 160f

R

Randolph, Theron, 1, 57–58
Range hood fans, 89
Rashes, 50–51, 51f
Rats, experiment in, 22, 46, 54, 58
Rea, W. J., 2, 4, 34
Red cell cholinesterase (RChE), 46, 49, 50
Regular maintenance, notes concerning, 92, 93
Regulations
building and statutory, *see* Building and statutory regulations
chemical features of indoor pollutants and, 123–124
impact, 124, 124f
time course, 123, 123t, 124
formaldehyde, 81–82, 83–84
in Japan, 139–142
Act on Maintenance of Sanitation in Buildings, 140, 141t
Housing Quality Assurance Act, 139, 140t
indoor density guideline value, 141, 142t
Revised Building Standards Act, 140, 141, 141t
School Health and Safety Act, 139, 140t
statutory, for indoor air quality, *see* Statutory regulations

TVOC concentration, 148–149
elimination period, enforcement of, 149
information sharing, 149
Repellents, 110–111
Research, on mechanism of CS, 23–26
detoxification, 23–25, 24f
nervous sensitivity acquisition, 25–26, 25f
Respiratory function, in CS patients, 34
Response(s)
countermeasures against
mitigation of SHS, 161–162, 161f
SHS, 162–163, 162f, 163f
dose–response curve, 18, 21, 22f, 23, 23f
environmental stressors, 57–58, 58f
Restrictions
concerning attic and adjacent concealed spaces, 84
on interior finishing, 83
Results, of investigation
field survey, SHS and, 153–156
chemical substance concentration measurements, 153, 154t
formaldehyde concentration and air change rate, 155, 155f
formaldehyde concentration and air-tightness measurement, 153, 155f
QEESI® questionnaire, 156, 156f
long-term observations, 159–161
chemical substance concentration and building age, 159, 159f, 160f
occupants' symptoms, yearly changes of, 160–161, 160f, 161f
measurement
air-tightness of houses, 100–102, 100f, 101f
ventilation rate, 102–103, 102f
Retardants, flame, 111–112, 111f
Revised Building Standards Act, 140, 141, 141t, 142, 146, 168–169, 170, 173
Risk assessment, indoor pollutants, 118–122
defined, 118
exposure, 120–121, 121f
hazard, 118–120, 119f, 120t
HQ and MOE, 121–122
Runow, K., 14

S

Saccadic eye movement, 30
Saku disease, 6
Sanitation, in building, 85–87
background and current status, 85, 85f
management, guidelines for, 85–86, 86t
unfit ratio, trend of, 86–87, 86f
Sauna, 40
Schaake, Reinhard, 67
School environment, CS and, 12, 124, 169
School Health and Safety Act, 139, 140t
Secondary generated compounds, 112
Secondary pollutants, 148
Selective serotonin reuptake inhibitors (SSRIs), for CS, 50–51, 51f
Selenium, 42, 42t
Selye, Hans, 57–58
Semivolatile organic compounds (SVOCs), 70, 73, 74
concentration, integration measurement, 129
emission of, measurement, 132–133
indoor chemical substances, 70, 70t
microchamber method for, 73–75, 74f
Serotonin reuptake, SSRIs, 50–51, 51f
Sharing, information, 149
Sick-building/house syndrome (SBS/SHS), 69, 75
causes of, 94, 94t
chemical substance concentrations, relationship, 156, 157f
children with, 46, 47; *see also* Children
defined, 69
development of, 49–50, 50f
diagnosis of, 29, 47, 167–168
emission rate of chemical compounds, *see* Chemical compounds
indoor air pollutants, 113–114
indoor air pollution, 80–81
measures against, 80–81
indoor environments and occupants' health, investigation, 151–163
field survey, 151–157; *see also* Field survey
long-term observations, 157–163; *see also* Observations, long-term
in Japan, 3, 8, 69
overview, 3
reduction of ventilation rate, 80
social recognitions, 147
symptoms of, 113, 156, 156f, 166, 167
Sick-building syndrome, 69
Sick-school problems, 124, 169
Single-photon emission computed tomography (SPECT), 33, 43
Small-chamber method, 69, 70–71, 71f, 72, 76
Smoking, 112
Smooth pursuit eye movement, 30, 30f, 31f
Social aspects, CS, 4–5
Social recognitions, of indoor air pollution, 146–148
new and remodeled houses, 146–147
Housing Quality Assurance Act, 146
nonregulated substances, 146–147, 147f
SHS, 147
routine countermeasures, 147–148
air cleaners, 148
oxidative decomposition, 148
secondary pollutants, 148

Source intensity, on-site method for, 130
Spa, 40
Specific adaptation syndrome (SAS), Randolph, 57–59, 58f
Spreading, defined, 11
Standing ability, of patients with CS, 32
Statutory regulations, for indoor air quality, 83–88
  building materials containing formaldehyde, 83–84
    attic and adjacent concealed spaces, restrictions, 84
    interior finishing, restrictions on, 83
    ventilation, mandatory installation, 83–84
  pollution, amendment of Building Standard Law (2003), 87, 87f, 88f
  technical standards
    Act on Maintenance of Sanitation in Buildings, 85–87, 85f, 86f, 86t
    in amended Building Standard Law, 83
    in Housing Quality Assurance Act, 84
Stool analysis, 64–65
Stressors, environmental, 57–58, 58f
Styrene, 76
Subject compounds
  instant measurement, 130
  integration measurement, 128–129
Sumatriptan succinate, for CS, 50–51, 51f
Supplement, defined, 43
Survey(s)
  field, *see* Field survey
  measurement, on chemical substance concentrations, 97–100
    period and investigated houses, 97, 98f, 98t
    points and method, 97, 98t, 99–100, 99f
  questionnaire
    about health conditions, 153
    contents of, 152
Symptomatic therapy, for CS, 41
Symptoms
  allergies, 12
  chemical substance concentrations and
    countermeasures against mitigation of SHS, 161–162, 161f
    countermeasures against SHS, 162–163, 162f, 163f
    field survey, 156–157, 158t
  children, CS in, 45–46
  CS, 3–4, 4t
  of damp building, 172
  ES, 13–14
  formaldehyde, cause, 167
  formaldehyde concentration and, 45
  occupants, yearly changes of, 160–161, 160f, 161f
  SBS/SHS, 113, 156, 156f, 166, 167
  TVOC concentration and, 46

T

Tachykinin, 26
Taurine, 40
Technical standards
  Act on Maintenance of Sanitation in Buildings, 85–87, 85f, 86f, 86t
  Building Standard Law, amended, 83
  Housing Quality Assurance Act, 84
Technology development, trends in, 90–91
  demand-controlled ventilation, 90
  hybrid ventilation systems, 91
  mechanical ventilation system
    with heat exchanger, 90–91
    utilizing low-power input motors, 90
Tenax® TA, 71, 129
Therapeutic diagnosis, 8
Thermotherapy, in detoxification, 40
Time course, of regulations, 123, 123t, 124
Tokyo Electric Power Corporation, 5
Toluene, 76, 110, 113, 124, 129
Total volatile organic compounds (TVOCs)
  air change rate and, relationship, 105, 105f
  air-tightness and, relationship, 103–104, 104f
  concentration, 146–147, 165–166, 170–171
    building age and, 159, 159f, 160f
    integration measurement, 128–129, 129f
    regulation of, 148–149
    symptoms and, 46
    temporal change of indoor, 146
  emission rate, 76
  guideline substances, concentrations of, 143
  instant measurement, 130
  personal exposure measurement method, 136
Toxicity
  assessment, 120, 122
  bronchial/pulmonary, 122
  building materials, 82
  chemical compound, 118–119
  education, 8
  metals, 60–61
  of organochlorine compounds, 50
  phthalates and parabens, exposure to, 61
  tests, 118–119
Toxins, 59–60
Tracking eye movement examination, 30, 30f, 31f
Tranquilizers, 41
Treatment, of CS, 37–44
  alternative therapy, 42–43
    alternative medicine, 43
    eastern medicine, 43
    prayer and meditation in Zen Buddhism, 42–43
  chemical substances, avoidance of, 37–39
    air, 38–39, 39f
    air purifiers, 38–39, 39f
    clothes, 39

food, 37
medicine, 39
water, 37–38
detoxification, 39–42
conjugation phase, 40
drug treatment, 40–42
EDTA, penicillamine, PAM, and others, 42
minerals, 41–42, 42t
spa, hot spring, bath, low-temperature sauna, 40
symptomatic therapy, 41
vitamins, 40–41
nutritional instruction, 43–44
Trichlorfon, 22
Tricresyl phosphate (TCP), 111
Triggers, 59
Trihalomethanes, 37
Triphenyl phosphate, 112
T-test, 156, 157
Tumor necrosis factor-alpha, 34

U

United States
CS and ES patients, 13, 14
CS cases in, 3–4, 4t
NIH in, 6
Urea melamine resin, 109
Urea resin, 109, 109f, 145

V

Ventilation, 97–105
measurement results, 100–103
air-tightness of houses, 100–102, 100f, 101f
rate, 102–103, 102f
measurement survey, 97–100
period and investigated houses, 97, 98f, 98t
points and method, 97, 98t, 99–100, 99f
rate, on-site method for measuring, 130
Ventilation fans, 38
common, 89
duct, 89
in HVAC systems, 90
range hood, 89
Ventilation system
demand-controlled, 90
equipment in large buildings, 93–95
centrally controlled HVAC systems, 94–95
in designated buildings, 93–94, 94t
installation, 83–84
interior finishing, restrictions on, 83
mandatory installation, 83–84
restrictions concerning attic and adjacent concealed spaces, 84
planning and operation, 91–93
concerning regular maintenance, notes, 92, 93
construction, notes on, 92
path, 91–92
pressure drops, notes for calculating, 92
steps for, 93, 93f
rate measurements, 102–103, 102f, 130
strategies for building and statutory regulations, *see* Building and statutory regulations
types, 88–90, 88t, 89t
hybrid, 89, 89t, 91
mechanical, *see* Mechanical ventilation system
natural, 88, 89
Verbal intelligence quotient (VIQ), 47, 49
Very volatile (gaseous) organic compounds, 70, 70t
Vinyl chloride, 12
Visual contrast sensitivity, 33
Vitamin B_{12}, 41, 60
Vitamin C, 40–41
Vitamin E, 41
Vitamins, for CS, 40–41, 41f
VOCs, *see* Volatile organic compounds (VOCs)
Volatile organic compounds (VOCs), 70, 71, 73, 76, 77
composition of, 143
concentration, integration measurement, 128–129, 129f
guideline substances, concentrations of, 143, 144t
high concentration substances, 143
indoor chemical substances, 70, 70t
indoor concentrations, in Japan, 114–118, 115t–117t
labeling, 76, 76t
measurement, 97
from paint, 110
SVOCs, *see* Semivolatile organic compounds (SVOCs)
TVOCs, *see* Total volatile organic compounds (TVOCs)
VVOCs, 70, 70t
VVOCs, 70, 70t

W

Water, avoidance of chemical substances, 37–38
Water-based paint and ink, 145
Wechsler Intelligence Scale for Children-Third Edition (WISC-III) tests, 47, 48, 49f

Wilcoxon rank sum test, 156, 157
Wolfhagen, Germany, 66–67, 67f
Wooden boards, artificial, 107, 109–110, 109f

X

Xenobiotics, defined, 40
Xenon, 18
Xylene, 76, 110, 113

Y

Yanagisawa sensor, 133, 134, 134f
Yoga, 42–43

Z

Zen Buddhism, prayer and meditation in, 42–43
Zinc, 42, 42t
Zonulin, 64